ANALYTICAL CHEMISTRY AND MICROCHEMISTRY

GAS CHROMATOGRAPHY

HISTORY, METHODS AND APPLICATIONS

ANALYTICAL CHEMISTRY AND MICROCHEMISTRY

Additional books and e-books in this series can be found on Nova's website under the Series tab.

ANALYTICAL CHEMISTRY AND MICROCHEMISTRY

GAS CHROMATOGRAPHY

HISTORY, METHODS AND APPLICATIONS

PERCY HENRICHON
EDITOR

nova science publishers
New York

Copyright © 2020 by Nova Science Publishers, Inc.

All rights reserved. No part of this book may be reproduced, stored in a retrieval system or transmitted in any form or by any means: electronic, electrostatic, magnetic, tape, mechanical photocopying, recording or otherwise without the written permission of the Publisher.

We have partnered with Copyright Clearance Center to make it easy for you to obtain permissions to reuse content from this publication. Simply navigate to this publication's page on Nova's website and locate the "Get Permission" button below the title description. This button is linked directly to the title's permission page on copyright.com. Alternatively, you can visit copyright.com and search by title, ISBN, or ISSN.

For further questions about using the service on copyright.com, please contact:
Copyright Clearance Center
Phone: +1-(978) 750-8400   Fax: +1-(978) 750-4470   E-mail: info@copyright.com.

NOTICE TO THE READER

The Publisher has taken reasonable care in the preparation of this book, but makes no expressed or implied warranty of any kind and assumes no responsibility for any errors or omissions. No liability is assumed for incidental or consequential damages in connection with or arising out of information contained in this book. The Publisher shall not be liable for any special, consequential, or exemplary damages resulting, in whole or in part, from the readers' use of, or reliance upon, this material. Any parts of this book based on government reports are so indicated and copyright is claimed for those parts to the extent applicable to compilations of such works.

Independent verification should be sought for any data, advice or recommendations contained in this book. In addition, no responsibility is assumed by the Publisher for any injury and/or damage to persons or property arising from any methods, products, instructions, ideas or otherwise contained in this publication.

This publication is designed to provide accurate and authoritative information with regard to the subject matter covered herein. It is sold with the clear understanding that the Publisher is not engaged in rendering legal or any other professional services. If legal or any other expert assistance is required, the services of a competent person should be sought. FROM A DECLARATION OF PARTICIPANTS JOINTLY ADOPTED BY A COMMITTEE OF THE AMERICAN BAR ASSOCIATION AND A COMMITTEE OF PUBLISHERS.

Additional color graphics may be available in the e-book version of this book.

Library of Congress Cataloging-in-Publication Data

ISBN: 978-1-53617-350-5

Published by Nova Science Publishers, Inc. † New York

CONTENTS

Preface   vii

Chapter 1   The Role of Gas Chromatography in Clinical and Forensic Toxicology   1
Ana Y. Simão, Mónica Antunes, Joana Gonçalves, Sofia Soares, Teresa Castro, Tiago Rosado, Débora Caramelo, Mário Barroso, André R. T. S. Araújo, Jesus Rodilla and Eugenia Gallardo

Chapter 2   Application of Sensor Gas Chromatography in Forensic Medicine   127
Hiroshi Kinoshita, Naoko Tanaka, Mostofa Jamal, Asuka Ito, Mitsuru Kumihashi, Tadayoshi Yamashita, Shoji Kimura, Yasuhiko Kimura, Kunihiko Tsutsui, Shuji Matsubara and Kiyoshi Ameno

Chapter 3   Trends of Gas Chromatography-Mass Spectrometry Techniques in Food Authentication   141
Oscar Núñez

Chapter 4	Gas Chromatography–Mass Spectrometry Analysis of Sugarcane Vinasse *Mohamed A. Fagier and Mona O. Abdalrhman*	**173**
Index		**193**

PREFACE

Gas Chromatography: History, Methods and Applications focuses on the main applications of gas chromatography in clinical and forensic toxicology, mainly in the determination of drugs of abuse including the new psychoactive substances in several types of biological matrices.

The authors go on to investigated the analysis of gaseous or volatile substances using sensor gas chromatography equipped with a semiconductor gas sensor detector. The simplicity, ease of handling, and high sensitivity of this method allow results to be obtained rapidly, which may provide valuable information for forensic diagnosis.

This compilation addresses the way in which food adulteration practices are potentially harmful to human health and so food safety and authenticity constitute an important issue in food chemistry. The chemical composition of foodstuffs is an excellent indicator of quality, origin, authenticity and/or adulteration.

The concluding study aims to determine the organic compounds of vinasse through gas chromatography-mass spectrometry GC-MS. Vinasse is a byproduct of ethanol and poses long-term risk to public health because of its persistent and toxic nature.

Chapter 1 - In the last two decades, without a doubt, liquid chromatography coupled to mass spectrometry has gained more relevance in the toxicology field, slowly and steadily replacing gas chromatography.

However, both instrumental techniques complement each other in routine laboratory analysis, and the development of mass spectrometry systems such as high-resolution mass spectrometry, time of flight or even orbitrap detectors, and two-dimensional chromatography systems or portable GC analysers as well, has definitely contributed to the reappearance of GC-based procedures.

The development of miniaturized systems that at the same time allow direct coupling to chromatographers has further increased the versatility of this instrumental technique. In fact, there are more and more publications in which these techniques are used for biological samples analysis, namely in what concerns the so-called alternative specimens, for instance oral fluid, hair, sweat and exhaled breath.

This chapter will focus on the main applications of gas chromatography in clinical and forensic toxicology, mainly in the determination of drugs of abuse including the new psychoactive substances in several types of biological matrices.

Some examples of the applications of this instrumental technique, as well as recent advances will be included, namely regarding hyphenated systems.

Chapter 2 - Gas chromatography is widely used for toxicological analyses in forensics for various kinds of gaseous or volatile substances, such as ethanol, carbon monoxide, hydrogen, cyanide. The present study investigated the analysis of these substances using sensor gas chromatography equipped with a semiconductor gas sensor detector. The simplicity, ease of handling, and high sensitivity of this method allow results to be obtained rapidly, which may provide valuable information for forensic diagnosis.

Chapter 3 - Food adulteration practices are potentially harmful to human health and so food safety and authenticity constitute an important issue in food chemistry. The chemical composition of foodstuffs is an excellent indicator of quality, origin, authenticity and/or adulteration. In general, food adulteration is carried out to increase volume, to mask the presence of inferior quality components, and to replace the authentic substances for the seller's economic gain. For instance, a common fraud is the employment of a cheaper similar ingredient, which the consumer has difficulty recognizing and which is difficult to detect by current analytical methodologies. For example, fruit-processed products are common adulterated by addition of water, sugars, fruits of inferior commercial value, secondary extracts of fruits and colors, etc. In other cases, the fraud is related to an incorrect labelling of the food product, for example, when refined or lower quality olive oils are labeled as extra-virgin olive oils. Thus, the development of analytical methodologies to achieve food authentication and to identify food frauds is required.

Gas chromatography coupled to mass spectrometry (GC-MS) is nowadays one of the most employed techniques to address food authenticity issues especially by the non-targeted fingerprinting of the food volatolome (volatile metabolites present in the foodstuffs). Targeted strategies by focusing either in the determination of specific food biomarkers or by the profiling of selected families of chemical compounds such as fatty acids are also employed within GC methodologies for authentication purposes. Compound-specific isotope analysis by isotope ratio mass spectrometry (IRMS) following the on-line combustion of compounds separated by GC has also become a method of choice in the authenticity control of foodstuffs based on the measurement of the isotope distribution at natural abundance level. In addition, multidimensional gas chromatography coupled with mass spectrometry has become also a powerful tool in food analysis, being also employed in food authentication problems. In this chapter, the role of GC-MS techniques for food authentication and the identification and prevention of frauds will be addressed. Coverage of all kind of applications is beyond the scope of the present contribution, so the present chapter will focus on the most relevant applications published in the last years.

Chapter 4 - Structural analysis of wastewater is one of the important issue in wastewater treatment plant to select suitable and effective methods of treatments. Vinasse is a byproduct of ethanol and poses long-term risk to public health because of their persistent and toxic nature. Vinasse as distillery wastewater is considered as complex matrices, therefore, this study, aimed to find the organic compounds of vinasse by using gas chromatography- mass spectrometry GC-MS.

The solvent extraction method was used for sample preparation. Hexane and dichloromethane (DCM) were used as solvent. About fifteen organic compounds were detected and confirmed, twelve of them were identified and confirmed by hexane extraction, while only three compounds were identified and confirmed by DCM extraction. In general, the most abundant phenolic compounds were: 4-ethyl-3-methoxy phenol, 2,6- dimethoxy phenol, 4-allyl-2,6-dimethoxy phenol and (3,4-dimethoxyphenoxy) trimethylsilane.

In: Gas Chromatography
Editor: Percy Henrichon

ISBN: 978-1-53617-350-5
© 2020 Nova Science Publishers, Inc.

Chapter 1

THE ROLE OF GAS CHROMATOGRAPHY IN CLINICAL AND FORENSIC TOXICOLOGY

Ana Y. Simão[1,2,], Mónica Antunes[3,*],
Joana Gonçalves[1,2,*], Sofia Soares[1,2,*], Teresa Castro[2,*],
Tiago Rosado[1,2,*], Débora Caramelo[1,2,*],
Mário Barroso[3*], PhD, André R. T. S. Araújo[4,5,6,*], PhD,
Jesus Rodilla[7,8], PhD and Eugenia Gallardo[1,2,†], PhD*

[1]Centro de Investigação em Ciências da Saúde,
Universidade da Beira Interior (CICS-UBI), Covilhã, Portugal
[2]Laboratório de Fármaco-Toxicologia, Ubimedical,
Universidade da Beira Interior, Covilhã, Portugal
[3]Serviço de Química e Toxicologia Forenses, Instituto Nacional
de Medicina Legal e Ciências Forenses I. P., Lisboa, Portugal
[4]Instituto Politécnico da Guarda (IPG), Escola Superior de Saúde,
Guarda, Portugal

---

* These authors contributed equally to the work.
† Corresponding Author's Email: egallardo@fcsaude.ubi.pt.

[5]Unidade de Investigação para o Desenvolvimento do Interior do Instituto Politécnico da Guarda (UDI/IPG), Guarda, Portugal
[6]Laboratório de Química Aplicada (LAQV, Requimte), Departamento de Ciências Químicas, Faculdade de Farmácia, Universidade do Porto, Porto, Portugal
[7]Materiais Fibrosos e Tecnologias Ambientais – FibEnTech, Departamento de Química, Universidade da Beira Interior, Covilhã, Portugal
[8]Departamento de Química, Universidade da Beira Interior, Covilhã, Portugal

Abstract

In the last two decades, without a doubt, liquid chromatography coupled to mass spectrometry has gained more relevance in the toxicology field, slowly and steadily replacing gas chromatography. However, both instrumental techniques complement each other in routine laboratory analysis, and the development of mass spectrometry systems such as high-resolution mass spectrometry, time of flight or even orbitrap detectors, and two-dimensional chromatography systems or portable GC analysers as well, has definitely contributed to the reappearance of GC-based procedures.

The development of miniaturized systems that at the same time allow direct coupling to chromatographers has further increased the versatility of this instrumental technique. In fact, there are more and more publications in which these techniques are used for biological samples analysis, namely in what concerns the so-called alternative specimens, for instance oral fluid, hair, sweat and exhaled breath.

This chapter will focus on the main applications of gas chromatography in clinical and forensic toxicology, mainly in the determination of drugs of abuse including the new psychoactive substances in several types of biological matrices.

Some examples of the applications of this instrumental technique, as well as recent advances will be included, namely regarding hyphenated systems.

Keywords: gas chromatography, forensic toxicology, clinical toxicology, biological specimens

INTRODUCTION

The first on-line coupling of gas chromatography to a mass spectrometer was reported in 1959, and its application in the several fields of analytical sciences has been increasing ever since (Hites 2016). There is no doubt that the improvements of this technique, such as increases in its sensitivity via coupling to several new detectors, namely tandem mass spectrometry (MS-MS) and time of flight (TOF) or improvements in analyte resolution via comprehensive two-dimensional gas chromatographic systems (GCxGC or 2D-GC), have contributed to the fact that this is one of the most versatile and used analytical equipments worldwide (Kyle 2017).

The event of coupling liquid chromatography systems to mass spectrometers (LC/MS) has lead to their widespread use by researchers and to the development of new instruments and applications that use this technique (Gallardo, Barroso, and Queiroz 2009b); however, both gas and liquid chromatographic systems play an important role in laboratories nowadays, and may be seen as complimentary in what concerns sample analysis.

We are constantly surrounded by chemicals, and as such our exposure to them may occur at home, work or environment. The great complexity of the toxic agents requires the use of sophisticated tools and analytical techniques to evaluate adequately toxic exposure. These evaluations usually involve qualitative and quantitative analysis to identify the toxic compound(s) and determine the respective amount in order to try to explain the oberved toxic syndromes characteristic of several classes of compounds (Mbughuni, Jannetto, and Langman 2016). Identifying the source of toxic exposure is very important as well.

However, laboratorial analyses aim at detecting and confirming the presence of a certain substance, and also provide prognosis in those situations where the results are capable of aiding patient management. The order of operation in toxicological analyses is as follows: extraction, purification, detection and quantification (whenever possible).

With the improvements of MS, GC and LC technologies in the second half of the twentieth century, sophisticated methods started to appear in modern toxicology laboratories for separation and detection, originating new and sensitive analytical applications, firstly developed in research scenarios but steadily implemented in routine laboratories. Mass spectrometry is undoubtedly the most used detection system in modern clinical and forensic toxicology (M. Prata et al. 2019; Malaca et al. 2019; Tiago Rosado, Gonçalves, Margalho, et al. 2017; Tiago Rosado, Oppolzer, et al. 2018; T. Rosado et al. 2017; M. Barroso, Gallardo, and Queiroz 2009; Caramelo et al. 2019; M. Barroso et al. 2005; Gallardo et al. 2006b, 2006a, 2006c; Amaral et al. 2010; Soares et al. 2019), only surpassed by flame ionization detection (FID) usually used for ethanol and other alcohols determination.

This chapter will be an approach on the use of gas chromatography systems, namely in the main applications in clinical and forensic toxicology (particularly concerning the determination of drugs of abuse, including new psychoactive substances) in several types of biological specimens.

DETERMINATION OF OPIATES AND OPIOIDS

Several methods have been developed and described to detect opiates in biological specimens, using thin-layer chromatography (Yeh 1973; Jain et al. 1975), immunoassays (Segura et al. 1999), gas chromatography and liquid chromatography coupled to several detectors (Tabarra et al. 2019; M. Prata et al. 2019; M. Barroso et al. 2011).

Immunoassays are not the most specific methods due to the substantial cross reactivity observed for opiates (Segura et al. 1999). For this reason, the presumptive positive samples need to be reanalysed and confirmed using chromatographic methods.

Nowadays, the advances made in analytical instrumentation, such as LC–MS/MS have proven to be an excellent choice regarding the simultaneous determination of naturally occurring opiates and their synthetic derivatives, namely hydrocodone and hydromorphone, in a wide range of biological matrices (Rana, Garg, and Singla 2014). However, this type of instrumentation is more expensive than traditional and widely used GC-MS systems, and for this reason it may be cost-prohibitive for many toxicology laboratories (Rana, Garg, and Singla 2014).

In fact, GC-MS has been recognized as the preferred method of choice for the determination of innumerous drugs. The adopted standard procedures (e.g., electron ionization, chemical ionization) involving the selected ion monitoring (SIM) approach present enough sensitivity to determine the amount of opiates in biological samples such as blood/serum, urine, oral fluid and hair, amongst others (Pichini et al. 1999). If coupled to MS/MS, sensitivity and selectivity can be further increased, allowing the determination of opiates at lower concentrations when matrix components create a chromatographic profile with many interferences (Pichini et al. 1999).

Nevertheless, problems are still found in GC–MS methods for simultaneous analysis of opiates like morphine and codeine that include interference from keto-opiates such as hydrocodone, hydromorphone, oxycodone and oxymorphone (Rana, Garg, and Singla 2014). These problems may be caused by incomplete derivatization, instability of the derivatives, poor chromatography, inefficient hydrolysis and extended run times (Rana, Garg, and Singla 2014). Many authors suggest procedures in order to improve the separation of these opiates (e.g., sequential derivatization, multiple ramp temperatures), but the available literature with reported alternatives is too extent.

When using GC-MS, the ionization mode can be relevant. The electron ionization mode (EI) is the most commonly used technique regarding drugs of abuse analysis (Y. H. Wu et al. 2008). The EI mode generates a number of fragment ions which provides more structural information, also allowing the identification of unknown analytes when their mass spectrum is compared with available libraries (Y. H. Wu et al. 2008). The chemical ionization mode (CI) represents a powerful tool in MS improving the selectivity and specificity.

The negative chemical ionization (NCI) has enhanced opiates determination in non-conventional matrices because it usually involves background noise reduction and subsequent increment in the signal-to-noise ratio (Y. H. Wu et al. 2008). However, positive chemical ionization (PCI) has also been successfully reported. Lowe et al., (Lowe et al. 2006) used methane as reagent gas and demonstrated that PCI afforded a two-fold sensitivity advantage over EI ionization when quantifying opiates amongst other drugs of abuse in human *postmortem* brain. Also, Wilkins et al. (Wilkins et al. 1995) reported PCI to determine codeine and its major metabolites in hair samples. These authors concluded that acetone reagent gas results in cleaner chromatograms than other reagent gases, less fragmentation under CI conditions, and strong, intense peaks corresponding to the protonated molecule. Other researchers use CI-MS for the determination of opiates with ammonia-methane (1:5) as the reactant gas (Drost et al. 1984).

However, the recent literature tends to focus more on the EI mode, generally at 70 eV. The determination of opiates usually applies 12- or 15m fused-silica capillary column constituted by nonpolar stationary phases of cross-linked dimethylsilicone, phenylmethylsilicone or 95% dimethyl-5% polysiloxane (Wasels and Belleville 1994; Fehn and Megges 1985; Bowie and Kirkpatrick 1989).

The oven temperatures may vary from maintaining it in isothermal mode at 230°C (Lora-Tamayo, Tena, and Tena 1987; ElSohly et al. 1988), to the most common temperature ramps, with initial temperatures comprehended between 50°C (B. H. Chen, Taylor, and Pappas 1990) and 160°C (Bermejo et al. 1992) and final temperatures from 240°C (Bermejo et al. 1992) to 280°C (Goldberger et al. 1991), the rate of increase being from 10°C/min (Inturrisi et al. 1984) to 50°C/min (B. H. Chen, Taylor, and Pappas 1990; Wasels and Belleville 1994). Regarding the carrier gas, helium is by far the most reported with flow rates ranging from 0.6 mL/min (Romolo et al. 2003) to 2.0 mL/min (Kovatsi et al. 2011). Nevertheless, hydrogen has also been reported (Rana, Garg, and Singla 2014).

It is a fact that some authors did not derivatize the opiates before chromatography (Caldwell and Challenger 1989; Masumoto et al. 1986; Wasels and Belleville 1994). However, the underivatized opiates reveal poor chromatographic properties (Wasels and Belleville 1994). The derivatization of the opiates is necessary to convert the polar hydroxyl groups into nonpolar derivatives which leads to chromatography improvement and an increase in sensitivity (B. H. Chen, Taylor, and Pappas 1990). The selection of derivatization method can be considered one crucial factor for the accuracy and precision of a GC-MS method (B. H. Chen, Taylor, and Pappas 1990). Several methods are available, such as acetylation, propionylation (propionic anhydride) and the formation of trimethylsilyl or perfluoroester derivatives. The number of derivatizing agents described in the literature may be relatively limited, but there is great variability in the experimental conditions (Wasels and Belleville 1994). When choosing a derivatizing agent it is important to consider whether the reagent will generate derivatives compatible with the chromatography of other compounds on the same column (Grinstead 1991). As example, trimethylsilyl derivatives should not be chromatographed on columns that will be also used to analyse underivatized drugs or derivatives produced from fluorinated anhydrides such as pentafluoro-propionic anhydride (PFPA) and heptafluorobutyric anhydride (HFBA) (Grinstead 1991). Moreover, the silyl and perfluoroacyl esters of opiates are readily hydrolysed, imposing limitations on the time of analysis (Melent'Ev 2004).

Chen et al., (B. H. Chen, Taylor, and Pappas 1990) made a comparison of derivatizing agents applied for the determination of codeine and morphine by GC-MS. The authors report that PFPA and HFBA derivatives of morphine and codeine resulted in poor spectra due to low abundances of secondary and tertiary ions. Grinstead (Grinstead 1991) presented similar results when using these two agents. In addition, trifluoroacetamide (MBTFA) appeared to be unsuitable for quantitative methods due to low stability observed for the internal standard adopted. The quantification of codeine and morphine using bis-trimethylsilyltrifluoroacetamide (BSTFA) showed a gradual decrease of the peak area ratio while the acetic anhydride derivative revealed no significant differences over a period of 24 hours (B.

H. Chen, Taylor, and Pappas 1990). Huang et al., (Huang, Andollo, and Hearn 1992) also tested different derivatizing agents for opiates identification in urine including trifluoroacetic anhydride (TFAA), PFPA, HFBA, BSTFA + 1% trimethylchlorosilane (TMCS), and acetic anhydride-pyridine.

These authors chose acetic anhydride derivatization because it yielded a single derivative for each opiate, with exception of hydrocodone, which is not acetylated (Huang, Andollo, and Hearn 1992). The fluoroacyl anhydrides were excluded due to the variable amounts of mono- and di-acyl derivatives generated, and were also excluded the silyl derivatives because were found less stable than the acetyl. Oxycodone and hydromorphone did not derivatize effectively with BSTFA + 1% TMCS (Huang, Andollo, and Hearn 1992). Balikova et al. (Balikova, Maresova, and Habrdova 2001) stated that N-methyl-N-(trimethylsilyl)trifluoroacetamide (MSTFA) use for quantitation of opiates by GC–MS is rather irreproducible for N-demethylated metabolites, and the reproducibility for other opiates remained unaffected by changing the silylation reagent. However, the use of MSTFA with ammonium iodide yielded much more reproducible results for these compounds.

Nowatzke et al., (Nowatzke et al. 1999) reported the distinction between eight opiate drugs in urine by GC-MS. The authors used BSTFA and found that opiates could be distinguished by analyses of TMS ether derivatives, and the identification of 6-keto opiates could be further confirmed by analysing methoxime (MO)-TMS derivatives (Nowatzke et al. 1999). Guthery et al. (Guthery et al. 2010) revealed that the sensitivity observed for opiate-type compounds was better with N-Methyl-N-tert-butyldimethyl-silyltrifluoroacetamide (MTBSTFA).

On the other hand, Melent'Ev (Melent'Ev 2004) investigated the determination of morphine and codeine in blood as their propionic esters reporting that these are more stable than the perfluoroacyl and silyl esters. Moreover, MSTFA + 5% TMCS has been reported as a good derivatizing agent for classic opiates, presenting good stability for at least 24 hours (T. Rosado et al. 2019; M. Prata et al. 2019).

Table 1. Some applications of GC to determine opiate drugs in clinical and forensic scenarios

Compounds	Sample	Volume/weight (mL or mg)	Extraction procedure	GC conditions	LOD	LOQ	Ref.
Tramadol, codeine, morphine, 6-acetylcodeine, 6-monoacetylmorphine and fentanyl	Hair	50	Digestion (methanol 65°C overnight) and MEPS (4 mg; 80% C8 and 20% SCX). Derivatization with MSTFA and 5% TMCS	GC-EI-MS/MS Capillary column (30m × 0.25-mm I.D., 0.25-μm film thickness) with 5% phenylmethylsiloxane (HP-5MS). The initial oven temperature was held at 90°C for 2 min, then raised to 300°C at 20°C/min (held for 3 min). The temperatures of the injection port and the transfer line were set at 240 and 280°C, respectively. The mass spectrometer operated with a filament current of 35μA and electron energy of 70 eV in the positive EI mode.	10 – 25 pg/mg	10 -25 pg/mg	(T. Rosado et al. 2019)
Morphine, codeine and 6-monoacetyl-morphine	Blood	0.25	PPT (acetonitrile) and microextraction by packed sorbent (4 mg; 80% C8 and 20% SCX). Derivatization with MSTFA and 5% TMCS	GC-EI-MS/MS Capillary column of fused silica (30 m ×0.25 mm, 0.25 μm i.d.) with 5% phenylmethylsiloxane (HP-5MS). The oven temperature started at 90 °C for 2 min, followed by an increase of 20 °C per minute until 300 °C, which was maintained for 3 min. The injector and detector temperatures were 220 °C and 280 °C, respectively; the source temperature was 230 °C. The mass spectrometer was operated with a filament current of 35 μA and electron energy 70 eV in the positive EI mode.	5 ng/mL	5 ng/mL	(M. Prata et al. 2019)
Morphine, codeine	Urine	1	SPE (Cerex Polycrom Clin II™ tubes). Derivatization with BSTFA and 1% TMCS	GC-EI-MS Capillary column DB-1 [15 m × 0.32 mm (i.d.), 0.25-μm-thick film]. The temperature program was: initial temperature, 150°C for 1.5 min; ramp at 20°C/min to 250°C; injection temperature, 250°C; and transfer line, 280°C.	n.a	300 ng/mL	(Smith et al. 2014)

Table 1. (Continued)

Compounds	Sample	Volume/weight (mL or mg)	Extraction procedure	GC conditions	LOD	LOQ	Ref.
Buprenorphine and Norbuprenorphine	Urine	5	SPE (GV-65C). Derivatization with BSTFA and 1% TMCS	GC-EI-MS Capillary column ZB-5MS (15 m x 250 μm ID, film thickness 0.25 μm). The oven program was as follows: initial temperature = 150°C, initial time = 0.50 min; ramp = 30°C/min; final temperature = 320°C; final time= 3.00 min.	10 ng/mL	10 ng/mL	(Gervais and Hobbs 2016)
Morphine, codeine, hydrocodone and hydromorphone	Urine	1	SPE (Clean Screen). Derivatization with BSTFA and 1% TMCS	GC-EI-MS Capillary column (10 m 0.15 mm, film thickness 0.12 μm). The oven temperature program was initiated at 150°C (held for 0.5 min), and ramped to 300°C at a rate of 40°C/min.. The injector temperatures was 250°C.	50 ng/mL	100 ng/mL	(Rana, Garg, and Singla 2014)
Codeine, morphine, and 6-monoacetyl-morphine	vitreous humor	0.1	Disposable pipette extraction (DPX CX-5). Derivatization with MSTFA.	GC-NPD Capillary column (5% phenyl and 95% methylpolysiloxane column (30mx0.32mm, 0.25 μm). The oven temperature was held at 110°C for 1min. It was then increased to 230°C at a rate of 10°C/min and was finally increased to 280°C at a rate of 40°C/min, where the temperature was held until the separation was completed. The temperatures of the injector port and the detector were set at 230 and 300°C, respectively.	0.16 - 1.25 μg/mL	0.49 - 3.79 μg/mL	(Kovatsi et al. 2011)
Morphine, 6-acetylmorphine, codeine, 6-acetylcodeine and tramadol	Hair	20	Digestion (methanol 65°C overnight) and solid phase extraction (Oasis® MCX). Derivatization	GC-EI-MS Capillary column (30 m × 0.25 mm I.D., 0.25 μm film thickness) with 5% phenylmethylsiloxane (HP-5MS) was used. Initial oven temperature was 90°C for 2 min, which was increased by 20°C min−1 to 300°C, held for 3 min. The temperatures of the injection port and detector were set at 220 and 280°C, respectively. The mass	50 pg/mg	50 pg/mg	(M. Barroso et al. 2010)

Compounds	Sample	Volume/weight (mL or mg)	Extraction procedure	GC conditions	LOD	LOQ	Ref.
			with MSTFA and 5% TMCS	spectrometer was operated with a filament current of 300 µA and electron energy of 70 eV in the EI mode.			
Codeine, morphine, and 6-acetylmorphine	Meconium	500	4 mL of methanol and SPE (BondElut solid-phase extraction). Derivatization with BSTFA and 1% TMCS	GC-EI-MS Capillary column HP-5MS (30-m × 250-µm; 0.25-µm film thickness). Initial oven temperature was 90°C (1 min) to 190°C (1 min) at 30°C/min, then to 260°C (4min) at 8°C/min and finally to 290°C (10 min). The injector was maintained at 240°C and operated for 2 min in splitless mode. The mass selective detector was kept at 300°C, the ion source at 250°C, and the quadrupole at 100°C. The mass analyser operated by EI (70eV).	5 – 10 ng/g	20 ng/g	(López et al. 2009)
Heroin, codeine, morphine, 6-acetylmorphine and 6-acetylcodeine	Sweat	patch	6 mL of 0.5 M sodium acetate buffer (pH 4.0) and SPE (Clean Screen® ZSDAU020). Derivatization with BSTFA and 1% TMCS	GC-EI-MS Capillary column HP-5MS capillary column (30 m × 0.32 mm i.d. × 0.25 µm film thickness). An initial oven temperature of 100°C was held for 0.5 min, followed by ramps of 25°C/min to 245°C, 2°C/min to 255°C and 30°C/min to a final temperature of 300°C for 0.7 min The temperatures of the quadrupole, ion source and mass selective detector interface were 150, 230 and 280°C, respectively. The injection port temperature was maintained at 200°C.	1.25 - 5 ng/patch	5 ng/patch	(Brunet et al. 2008)
6-monoacetylmorphine, morphine and codeine	Oral Fluid	1	SPE (Bond Elut Certify®). Derivatization with MSTFA	GC-EI-MS Methylsilicone capillary column (Ultra 1, 16.5 m × 0.2 mm i.d., 0.11 µm film thickness). The oven temperature was programmed at 70°C (2 min), followed by a 30°C/min ramp to 160°C, 5°C/min to 170°C, 20°C/min to 200°C, 10°C/min to 220°C and finally increased 30°C/min ramp to 300°C. The injector and the interface were operated at 280°C. The mass spectrometer was operated in EI mode at 70 eV.	0.9 - 2.2 ng/mL	2.9 - 6.6 ng/mL	(Pujadas et al. 2007)

Table 1. (Continued)

Compounds	Sample	Volume/weight (mL or mg)	Extraction procedure	GC conditions	LOD	LOQ	Ref.
6-monoace-tyl-morphine, morphine and codeine	Teeth	1000	Incubation with 2 mL 0.1 M HCl at 37°C for 18 h and LLE (3 mL of chloroform/ isopropanol (9:1)). Derivatization with BSTFA and 1%TMCS	GC-EI-MS Capillary column (HP-5MS, 30 m × 0.25 mm i.d, film thickness 0.25 μm). The oven temperature was programmed at 80°C for 1 min, increased to 230°C at 35°C/min, and then raised to 290°C at 10°C/min and held for 10 min. Split injection mode (15:1). The injection port, ion source, quadrupole and interface temperatures were: 260, 230, 150 and 280°C, respectively. EI mode was used.	2.0 - 2.5 ng/g	6.0 - 7.5 ng/g	(Pellegrini et al. 2006)
Morphine, codeine, 6-acetylmorphine	Brain tissue	100 – 150	2 M sodium acetate (pH 4.0) and SPE (Clean Screen® ZSDAU020). Derivatization with N-methyl-N-(tert-butyldimethylsilyl) trifluoroacetamide and 1% tert-butyldimethylchloro-silane and with BSTFA +1% TMCS	GC-PCI-MS Capillary column HP-1MS (30 m x 0.32 mm i.d., 0.25 μM film thickness). Initial column temperature of 70°C was held for 1.00 min, increased to 175°C at 30°/min, ramped to 250°C at 23°/min, and increased to a final temperature of 310°C at 18°/min that was held for 5.00 min. The MS was PCI mode with methane reactant gas at a flow control setting of 23%. MS interface, source, and quadrupole temperatures were 295, 250, and 150°C respectively.	50 ng/g	50 ng/g	(Lowe et al. 2006)

Legend: BSTFA: N,O-bis-(trimethylsilyl)-trifluoroacetamide; EI: Electron Ionization; GC-MS: Gas chromatography coupled mass spectrometry; GC-MS/MS: Gas chromatography coupled mass spectrometry in tandem; LLE: Liquid-liquid extraction; LOD: Limit of detection; LOQ: Limit of quantitation; MEPS: microextraction by packed sorbent; MSTFA: N-Methyl-N-trimethylsilyl-trifluoroacetamide; MTBSTFA: tert-butyldimethylchlorosilane; n.a: not available; PCI: Positive Chemical Ionization; PPT: Protein precipitation; SPE: Solid-phase extraction; SPME: solid-phase microextraction; TMCS: trimethylchlorosilane.

In the absence of the LOQ value, the lowest point of the calibration curve was considered.

Although, as previously mentioned, MS is the most adopted detection system for opiates, several others are reported in the literature. Silylated morphine has been detected with a FID. As alternative, nitrogenphosphorus detection (NPD), also known as flame thermionic detection (FTD), should provide greater selectivity and sensitivity towards morphine than that observed with FID, with only a small increase in expense (H. M. Lee et al. 1991). Vu-Duc, T. and Vernay, A. (Vu-Duc and Vernay 1990) presented a method for the simultaneous determination of 6-monoacetylmorphine (6-MAM), morphine and codeine in human urine by capillary GC associated to a nitrogen selective detector (NSD) and FID. The authors reported that sensitivity of NSD for 6-MAM was fivefold greater than that provided by the FID, however this was not consistent when the determination was performed in authentic samples due to the increase of background noise (Vu-Duc and Vernay 1990).

More recently, comprehensive two-dimensional gas chromatography (GCxGC or 2D-GC), has proven to be a powerful technique and very suitable for the separation of complex mixtures, when compared to conventional single dimensional GC (1D-GC) (Noorizadeh and Noorizadeh 2012). 2D-GC provides an added dimension by separating the sample components in two phases, also known as orthogonal separation, prior to detection (Guthery et al. 2010). Noorizadeh and Noorizadeh (Noorizadeh and Noorizadeh 2012) studied the quantitative structure-retention relationship (QSRR) of 69 opiate and sedative drugs against the 2D-GC retention time. The authors have successfully separated all target analytes firstly by their differences in boiling point using a conventional dimension low polarity column, and then by their polarity differences through the use of a higher polarity column (Noorizadeh and Noorizadeh 2012). Also, Guthery et al. (Guthery et al. 2010) performed a qualitative drug analysis that included codeine, morphine and 6-MAM in hair extracts by comprehensive 2D-GC coupled to TOF. The authors stated that the advantage of comprehensive 2D-GC separation is that analytes that co-elute in the first column may be efficiently resolved in the second, usually with different polarities (Guthery et al. 2010). This will result in a significant reduction of interfering peaks. In addition, the levels of noise caused by column and septum bleed and other contaminants can be virtually eliminated using 2D-GC (Guthery et al. 2010).

Some of the most recent applications of GC for the determination of opiates are presented in Table 1. The relevance of GC use in clinical and forensic toxicology analysis, specifically for this class of drugs, is shown by the large difference of matrices used. A good sample preparation coupled to optimal separation conditions on the chromatograph and a mass spectrometry detection is of outmost importance for a successful identification of different opiates

DETERMINATION OF COCAINE AND METABOLITES

Cocaine is one of the most popular stimulant drugs used worldwide, mainly exerting its effects on dopaminergic neurons. Its main metabolites are benzoylecgonine and ecgonine methyl ester. Both are inactive; the former is the most preponderant of the two and can be further metabolized to ecgonine. When cocaine is consumed with ethanol containing drinks, cocaethylene is obtained, while crack cocaine consumption originates anhydroecgonine methyl ester, a specific marker of this type of consumption. This derivative is thermally unstable, which can pose several difficulties in analytical methods (Feliu et al. 2015). A lot of different methodologies have been developed over the years in order to detect and quantify such compounds in several different types of biological matrices, such as blood, urine, hair, oral fluid, *postmortem* samples and others (Barroso, Gallardo, and Queiroz 2009). Still, there are a few items that should be considered prior method development: usually cocaine metabolites need a derivatization step in order to increase thermal stability and/or analyte volatility, as well as increasing molecular weight, which produces ions with bigger mass, allowing for a better specificity (Isenschmid, Levine, and Caplan 1988) (Rosado et al. 2017). Bioanalytical methods for cocaine and metabolites determination using GC as a main analytical technique will be reviewed.

Blood (plasma/serum) and urine samples are the most used biological specimens to document and report human exposure to drugs, and a number of reasons can be pointed out to this, for instance the ability to associate levels to the clinical symptoms in the case of the former and the amount of sample that is collected, allowing for more tests to be performed in the case of the latter (Barroso and Gallardo 2015). Several analytical methods have been developed and validated to determine cocaine in these specimens.

Lerch et al. (Lerch, Temme and Daldrup 2014) conducted a study comparing both an automated and a manual method. The authors quantified cocaine and its metabolite benzoylecgonine (and opioids and metabolites as well) from urine, blood (serum), and alternative matrices, namely brain and liver. The limits of detection (LOD) obtained for cocaine and benzoylecgonine were 1.1 and 9 ng/mL, respectively. Whereas the limits of quantification (LOQ) for the same compounds were 3.5 and 47 ng/mL, respectively. Although these limits are high when compared to others (Rosado et al. 2017; Fernández et al. 2019), the main goal of this work was achieved, and the automated method presented advantages for routine laboratory analysis, such as the possibility of using a large number of real samples, less prone to human error, method flexibility and suitability for several different matrices.

Ellefsen et al. (Ellefsen et al. 2015) performed a comparative study assessing the concentrations of cocaine and metabolites (benzoylecgonine, norcocaine and cocaethylene) using different approaches for analyte detection. On the one hand, cocaine and its metabolites were determined by using capillary dried blood spots (DBS) as extraction technique and LC coupled to high resolution mass spectrometry (HRMS).
On the other hand, whole blood was used to determine cocaine and benzoylecgonine using solid phase extraction (SPE) as extraction technique and 2D-GC-MS. Both techniques were studied at the same time and following controlled intravenous cocaine administration. The former allowed obtaining wider concentration ranges and lower LOQs. Also, this matrix proved to be less invasive and had a narrow capacity for adulteration.

Nonetheless, the variability regarding intra- and inter- day precision and accuracy was notorious. Furthermore, the coefficients of variation obtained were higher when compared to whole blood (for cocaine and benzoylecgonine), which accounts as a disadvantage for this technique and should be taken into consideration. Regarding the LOQs obtained when using whole blood as a matrix, the limits of quantification were as low as 1 ng/mL, except for norcocaine, which was not quantified in this matrix, since as the authors reported it did not achieve clinical relevance.

Using a different extraction technique, Rosado et al. (Rosado et al. 2017) have published a study using microextraction by packed sorbent (MEPS), which proved to be an advantageous approach due to the fact that less sample and solvent volumes were used, being a less time-consuming technique as well. By using just 200 μL of urine sample, researchers were able to achieve linearity and limits of detection and quantification as low as 25 ng/mL for all compounds under study.

Furthermore, real urine samples were successfully analysed, proving that the method can be successfully used in routine drug analysis in both clinical and forensic contexts. Additionally, the method was little time-consuming (15 min), highly sensitive and selective.

It is also possible to modify previously established methods in order to obtain better limits, recovery or even to improve one or more stages, such as extraction or derivatization steps. That was the case of the study published by Serafin et al. (Serafin et al. 2017); in this particular study the derivatization step formerly used by Stout et al. (Stout et al. 2002) was object of improvement by replacing the derivatization reagents, aiming at developing a more cost-effective method. The derivatization step was changed in two different ways: firstly, by using acetic anhydride (ethanoic anhydride), instead of PFPA, and secondly, by promoting a reaction between benzoylecgonine and 2,2,3,3,3-pentafluoropropanol (PFPOH), with no anhydride present.

As a result, no significant differences were observed when comparing both reagents, regarding the quantified concentration of benzoylecgonine. Still, acetic anhydride can act as a substitute for PFPA, which proved to be economically beneficial.

Most recently, Fernández et al. (Fernández et al. 2019), conducted a study where an optimized and fully validated analytical method was capable of analysing cocaine and eight metabolites in human urine. The quantification was operated by a GC/EI-MS and the analytes were extracted from the sample via a SPE technique consisting of a one-step extraction and using a small amount of sample (0.5 mL); also, the analytes suffered a derivatization step. The LOQs obtained were within the 2.5 to 10 ng/mL range. The method was considered remarkably sensitive and also costly-effective, considering the use of only one SPE cartridge and low volume of sample and solvent consumption. Moreover, it was considered a fast and rapid method to be applied to real samples, becoming useful in routine drug analysis.

Hair is one of the most used 'alternative' matrices, due to its capability to document human long-term exposure to drug abuse, since its detection window is much larger (from weeks to months, according to the length) than blood or urine (days) (Barroso, Gallardo, and Queiroz 2009; Gallardo and Queiroz 2008). The Society of Hair Testing (SoHT) (Society of Hair Testing 2019) proposes criteria for drug analysis, and in the case of cocaine it has been recommended to include in the chromatographic analysis cocaine and at least one of the following: benzoylecgonine, cocaethylene, norcocaine or ecgonine methyl ester.

Over more than 25 years ago, Möller et al. (Möller, Fey, and Rimbach 1992), developed a new method to identify and quantify cocaine and two of its metabolites (benzoylecgonine and ecgonine methyl ester) in twenty hair samples of miners who chewed coca leaves using GC-MS as an analytical tool.

By using small amounts of hair sample (around 10-30 mg) the authors were able to achieve LODs and LOQs in the range of ng/mg. Moreover, the results were able to gauge the time of consumption (4-11 months) and the method proved to be robust to determine these compounds in authentic hair specimens.

A few years later, Kintz et al. (Kintz and Mangin 1995) developed a different method to quantify cocaine and metabolites (benzoylecgonine, ecgonine methyl ester and cocaethylene) as well as opiates in human hair. Although the amount of hair sample used in this method was similar to that used by Möller et al. (Möller, Fey, and Rimbach 1992), Kintz et al. (Kintz and Mangin 1995) were able to obtain lower LODs (0.05; 0.20; 0.80 and 0.10 ng/mg for cocaine, benzoylecgonine, ecgonine methyl ester and cocaethylene, respectively).

López-Guarnido et al. (López-Guarnido et al. 2013) developed a method to quantify cocaine and also developed a new criterion to diminish the occurrence of false-negatives. The LODs obtained were as low as 0.01; 0.04 and 0.03 ng/mg for cocaine, benzoylecgonine and cocaethylene respectively, whereas the limit of quantification for all analytes was 0.2 ng/mg. Moreover, the method was considered robust and properly applicable to real hair samples, making it a useful tool in the forensic field. Also, the new criteria applied on this method were tested by the authors on previously published papers and they were able to minimize false positive results, which is important to investigate when working with hair samples, due to the possible external contamination.

In a different study, Aleksa et al. (Aleksa et al. 2012) developed a method to simultaneously quantify cocaine and metabolites (amongst other drugs), using GC-MS and two-step extraction, first using SPE, followed by solid phase microextraction (SPME) step after derivatization. ELISA (enzyme-linked immunosorbent assay) tests were used prior to the analytical method identification and detection of the analytes. The limits obtained were satisfying low, and in the range of ng/mg (i.e., 0.13 ng/mg for cocaine). Real samples quantification proved to be specific and sensitive and respected the guidelines used, allowing to say that this method can be applied to a routine forensic and clinic laboratory analysis.

Pego et al. (Pego et al. 2017) proposed a method to assess cocaine and metabolites in *postmortem* hair samples, using liquid phase microextraction (LPME) as a clean-up step prior to derivatization and GC-MS. The method was fully validated with LODs (ng/mg) for cocaine, anhydroecgonine methyl ester, cocaethylene, benzoylecgonine were 0.1; 0.4; 0.03 and 0.05, while LODs of 0.5 ng/mg (cocaine and anhydroecgonine methyl ester) and 0.05 ng/mg (cocaethylene and benzoylecgonine) were obtained. The method was successfully applied to real samples, proved to be selective, linear, and precise, which makes possible to apply as a fast routine analysis method (around 17 min). Also, the extraction procedure is considered as advantageous for being "green".

Cocaine and metabolites analysis in hair follows the pattern of other biological specimens, and as such extracts need to be derivatized prior to GC-MS. Care should be taken, however, in cases where minoxidil is present and derivatization MSTFA is used, since it yields an interfering peak at the same retention time and selected ions of cocaine, impairing its adequate confirmation (Zucchella et al. 2007).

Oral fluid is an easy attainable sample without the need of invasive collection procedures which can be collected under supervision, and this limits the possibility of adulteration or substitution (Barroso and Gallardo 2015; Drummer 2005; Gallardo, Barroso, and Queiroz 2009). The amount of sample available for analysis is usually small (less than 1 mL), and therefore extremely sensitive techniques are required, being MS mandatory. In addition, drug concentrations in this matrix can be correlated to blood/plasma concentrations, and eventually to the observed symptoms and degree of impairment as such. This sample thus becomes important and useful as a tool in the assessment of individuals suspected of driving under the influence of drugs.

In 2003, Cámpora et al. (Cámpora et al. 2003) conducted a study to quantify cocaine and its main metabolites ecgonine methyl ester and benzoylecgonine. The analytes were isolated from the sample by means of LLE and the LODs were 2.2; 0.9 and 0.2 ng/mL for cocaine, ecgonine methyl ester and benzoylecgonine, respectively. Due to the low polarity of some analytes, a derivatization step was performed. The authors analysed 48 samples using this method and 46 cases proved to be positive for at least one of the drugs, whereas 9 of the samples contained all three analytes.

Cognard et al. (Cognard, Bouchonnet, and Staub 2006) have used a previous developed method that quantified cocaine and metabolites in hair samples and applied it to oral fluid samples. The LODs ranged from 0.1 to

0.5 ng/mL (lower than those obtained by Cámpora et al. (Cámpora et al. 2003) and using half of the sample volume). The method was applied to real samples and proved to be applicable for the determination of all analytes under study, except for anhydroecgonine methyl ester. This last compound was only determined in a semi-quantitative fashion, considering the criteria applied to the study. Yet, the method is suitable to routine laboratory and forensic toxicological analysis.

As previously mentioned, oral fluid has become a popular tool to evaluate driving under the influence situations. Ellefsen et al. (Ellefsen et al. 2016) have performed a screening test by collecting 10 oral fluid samples with different commercially available devices after intravenous cocaine administration, followed by confirmation of the results through a fully validated 2D-GC-MS method. The aim of this study was to compare different oral fluid collecting devices and the authors concluded that even though there are slightly differences regarding $T_{1/2}$ of the compounds, these disparities are minor.

Child exposure to drugs studies can be performed non-invasively using placenta, amniotic fluid and umbilical cord (*in utero* exposure), as well as in breast milk. It is very important to understand that some samples have a more limited window of detection than others. Regarding *postmortem* analysis, fluid or tissue samples can be used, either in solid or liquid form.

Ripple et al. (Ripple et al. 1992) developed a method to detect cocaine and metabolites benzoylecgonine, ecgonine methyl ester and cocaethylene in amniotic fluid samples and three maternal samples. The sample volumes used ranged from 5 to 30 mL, obtaining for all analytes LODs and LOQs as low as 5 and 10 ng/mL, respectively. The method was considered linear and with good precision and accuracy, also it was sensitive to be applied to real samples.

Vitreous humor can be used also for drug determination, and this was the aim of a study performed by Peres et al. (Peres et al. 2014) using SPE and GC-MS for the detection of cocaine and metabolites (cocaethylene, benzoylecgonine, anhydroecgonine methyl ester) and several other drugs. The method was validated and the LODs obtained were as low as 1.0 ng/mL (anhydroecgonine methyl ester) and 2.0 ng/mL for all other analytes.

Table 2. Review of the GC methods developed for cocaine and metabolites determination in biological specimens

Compounds	Sample	Volume/ weight (mL or mg)	Extraction procedure	GC conditions	LOD	LOQ	Ref.
Cocaine and benzoylecgonine and opioids and derivatives	Urine, Blood, serum, brain and kidney (these last two, native and in lyophilized form)	0.6 (liquid samples); 600 (solid samples)	Automated SPE (RapidTrace SPE Workstation). Derivatization with isooctane/MSTFA (19/1 v/v) (manual derivatization) and isooctane/pyridine/MSTFA (14/5/1 v/v) (automated derivatization)	GC-MS Capillary column (30m × 0.25-mm I.D., 0.25-μm film thickness) with 5% phenylmethylsiloxane (HP-5MS). The initial oven temperature was held at 140°C for 1 min, followed by an increase of 120°C per minute until 225°C, which was maintained for 5.29 min, then raised to 300°C, maintained for 5.2 min. The mass spectrometer was operated with an electron energy of 70 eV in the selected ion monitoring mode.	1.1 - 9 ng/mL	3.5 - 47 ng/mL	(Lerch, Temme, and Daldrup 2014)
Cocaine, benzoylecgonine, norcocaine	Blood	0.3 (DBS); 0.25 (Whole blood)	DBS: (3 mm diameter disc, followed by SPE (SOLA™ CX); Whole blood: SPE (UCT Clean Screen). Derivatization with MTBSTFA and 1% t-butyl-dimethylchlorosilane	2D-GC-MS Capillary column (DBS-1MS,15 m × 0.25 mm, 0.25 μm and (ZB-50, 30 m × 0.32 mm, 0.25 μm) The initial oven temperature was held at 150°C for 0.5 min, followed by an increase of 30°C per minute until 290°C for 0.5 min (ramp 1), followed by an increase of 75°C per minute until 180°C for 1.5 min (ramp 2), the finally raising 30°C per minute until 290°C for 1 min (ramp 3). Post ramp: 40°C/min until 320°C for 4 min. The mass spectrometer was operated in EI mode	0.5 – 1 ng/mL	1 ng/mL	(Ellefsen et al. 2015)

Table 2. (Continued)

Compounds	Sample	Volume/ weight (mL or mg)	Extraction procedure	GC conditions	LOD	LOQ	Ref.
Cocaine, ecgonine methylester and benzoylecgonine	Urine	0.2	100 μL of 0.1 mM potassium phosphate buffer and spiked with 20 μL of the IS working solution, followed by MEPS (4 mg; 80% C8 and 20% SCX). Derivatization with MSTFA with 5% TMCS	GC-MS Capillary column (30m × 0.25-mm I.D., 0.25-μm film thickness) with 5% phenylmethylsiloxane (HP-5MS). The initial oven temperature was held at 90°C for 2 min, then raised to 300°C, maintained for 2 min. The temperatures of the injection port and the transfer line were set at 220°C and 280°C, respectively. The mass spectrometer was operated with a filament current of 35μA and electron energy of 70 eV in the positive EI mode.	25 ng/mL	25 ng/mL	(Tiago Rosado, Gonçalves, Margalho, et al. 2017)
Benzoylecgonine	Urine	2000	SPE (Speedisk 48 ™). Derivatization with PFP-OH and PFPA	GC-MS Capillary column (12m × 0.25-mm I.D., 0.25-μm film thickness) with 5% phenylmethylsiloxane (HP-5MS). The oven temperature was maintained isothermally between 225°C and 245°C. The injection port was maintained 20°C above the initial oven temperature and the transfer line temperature was maintained at 250°C.	n.a	n.a	(Serafin et al. 2017)
Cocaine, ecgonine, benzoylecgonine ecgonine methylester,	Urine	0.5	SPE (Clean Screen ®). Derivatization with PFPA and HFIP	GC-MS Capillary column (30m × 0.25-mm I.D., 0.25-μm film thickness) with 5% phenylmethylsiloxane (HP-5MS). The initial oven temperature was held at 150°C for 3 min, followed by an increase of 25°C per minute until 170°C, followed by an increase of 5°C per minute until	n.a	2.5 - 10 ng/mL	(Fernández et al. 2019)

Compounds	Sample	Volume/ weight (mL or mg)	Extraction procedure	GC conditions	LOD	LOQ	Ref.
cocaethylene, norcocaine, norcocaethylene, norbenzoylecgonine and m-hydroxybenzoylecgonine				240°C, then finally increasing 25°C per minute to 280°C, which was maintained for 2 min. The temperatures of the injection port and the transfer line were set at 250°C and 280°C, respectively. The mass spectrometer was operated with a filament current of 300 mA and electron energy of 70 eV in the positive EI mode.			
Cocaine, benzoylecgonine and ecgonine methylester	Hair	10–30	Wash (warm water for 5 min and acetone for 1 min). Hydrolyzation with β-glucuronidase-arylsulfatase for 2 h at 40°C, followed by centrifugation. SPE (Chromabond ® C18, 200 mg, 3 mL). Derivatization with PFPA and PFPOH	GC/MS HP-Ultra 2 Capillary column (12m × 0.2-mm I.D., 0.33-μm film thickness) with crosslinked 5% phenylmethylsiloxane (HP-5MS). The initial oven temperature was held at 70°C for 3 min, followed by an increase of 15°C per minute until 180°C, followed by an increase of 5°C per minute until 240°C, and then increasing 30°C per minute to 300°C, which was maintained for 5 min. The temperatures of the injection port and the transfer line were set at 260°C and 280°C, respectively. The mass spectrometer was operated with an electron energy of 70 eV in the EI mode.	0.1 – 1 ng/mg	0.5 ng/mg	(Möller, Fey, and Rimbach 1992)

Table 2. (Continued)

Compounds	Sample	Volume/ weight (mL or mg)	Extraction procedure	GC conditions	LOD	LOQ	Ref.
cocaine, benzoylecgonine, ecgonine methyl ester, cocaethylene	Hair	More than 30 (ideally 50)	Neutralization (1 mL of NaOH 0.1M) LLE(10 mL chloroform-isopropanol-n-heptane (50:17:33, v/v) and re-extraction (2 mL phosphate buffer, 1M NaOH and 5 mL chloroform). Derivatization with BSTFA and 1% TMCS	GC/MS Capillary column (12m × 0.22-mm I.D.) SGE with 5% phenyl-95%methylsiloxane (BP-5). The initial oven temperature was held at 60°C, followed by an increase of 30°C per minute until 310°C, which was maintained for 3 min. The temperatures of the injection port and the ion source were set at 260°C and 210-220°C, respectively. The ion-trap detector was operated with an electron energy of 70 eV and the electron multiplier voltage was set at 1650 V.	0.05 - 0.80 ng/mg	0.4 ng/mg	(Kintz and Mangin 1995)
cocaine, benzoylecgonine, cocaethylene	Hair	50	SPE (Oasis HLB®)	GC/MS Capillary column 5% phenylmethylsiloxane (HP-5MS) (30m × 0.25-mm I.D., 0.25-μm film thickness). The initial oven temperature was held at 90°C for 1 min, followed by an increase of 25°C per minute until 250°C, maintained for 5 min, finally followed by an increase of 1°C per minute until 255°C. The mass spectrometer was operated with an electron energy of 70 eV in the EI mode.	0.01 – 0.04 ng/mg	0.2 ng/mg	(López-Guarnido et al. 2013)
cocaine, benzoylecgonine,	Hair	10	Wash with dichloromethane for 1 min, followed by digestion (methanol at 56°C and during 18h). The SPE	GC/MS Capillary column (20m × 0.25-mm I.D., 0.25-μm film thickness). The initial oven temperature was held at 60°C for 1 min, followed by an increase of	0.13 ng/mg	0.4 ng/mg	(Aleksa et al. 2012)

Compounds	Sample	Volume/weight (mL or mg)	Extraction procedure	GC conditions	LOD	LOQ	Ref.
norcocaine, cocaethylene			(Oasis HCX® 60 mg 3 cm³ was performed overnight, followed by HS-SPME (pre-incubation for 5 min with 1 min agitations (250 rpm) and 15 s stop intervals, desorption time was 10 min; T= 80°C). Derivatization with BSTFA and MSTFA with 1% TMCS	15°C per minute until 222°C, then followed by an increase of 5°C per minute until 255°C, finally followed by raise of 30°C per minute until 300°C, maintained for 5 min. The temperatures of the injection port and the transfer line were set at 240°C and 300°C, respectively.			
Cocaine, anhydroecgonine methyl ester, cocaethylene and benzoylecgonine	Hair	50	Wash with detergent, water, dicholoromethane (15min t 37°C). Digestion with 2 mL of methanol (50°C for 18h). Derivatization with 100 μL of acetonitrile, 2.0 μL of pyridine and 2.0 μL of buthylchloroformate. LPME (organic phase composed by dihexyl ether and acceptor phase composed by 0.05M of HCl)	GC/MS Capillary column (30m × 0.25-mm I.D., 0.25-μm film thickness) with 5% phenylmethylsiloxane (HP-5MS). The initial oven temperature was held at 90°C for 1 min, followed by an increase of 15°C per minute until 250°C, maintained for 2 min, finally followed by an increase of 25°C per minute until 280°C, maintained for 2 minutes. The temperatures of the injection port and the transfer line were set at 250°C and 280°C, respectively. The mass spectrometer was operated with an electron energy of 70 eV in the EI mode.	0.03 - 0.4 ng/mg	0.5 ng/mg	(Pego et al. 2017)

Table 2. (Continued)

Compounds	Sample	Volume/weight (mL or mg)	Extraction procedure	GC conditions	LOD	LOQ	Ref.
Cocaine, ecgonine methylester and benzoylecgonine	Oral fluid	1000	LLE (Toxitube A®). Derivatization with BSTFA/TMCS	GC-PCI-MS Capillary column (12m × 0.33-mm I.D, 0.33-μm film thickness) with 5% phenylmethylsiloxane (HP-5MS). The initial oven temperature was held at 90°C for 1 min, followed by an increase of 20°C per minute until 180°C, followed by an increase of 5°C per minute until 240°C, the finally raising 30°C per minute until 290°C. The temperatures of the injection port and the transfer line were set at 250°C and 320°C, respectively. The mass spectrometer was operated with an electron energy of 70 eV in the PCI mode.	0.2 - 2.2 ng/mL	0.8 – 7.4 ng/mL	(Cámpora et al. 2003)
Cocaine, anhydroecgonine methylester, ecgonine methyl ester and cocaethylene	Oral fluid	0.5	Automated SPE (ASPEC apparatus) and HCX Isolute cartridges (130 mg).	GC-IT-MS/MS Capillary column (15m × 0.25-mm I.D, 0.25-μm film thickness) with 5% phenylmethylsiloxane (HP-5MS) connected to an inert retention gap of 1.5mm × 0.53mm I.D. The initial oven temperature was held at 75°C for 1 min, followed by an increase of 15°C per minute until 170°C, followed by an increase of 5°C per minute until 210°C, the finally raising 30°C per minute until 310°C. The temperature of the injection port was set at 75°C for 1 min then increased at 50°C per minute to 280, maintained for 1.40 min and the temperature of the transfer line was set at 290°C. The mass spectrometer was operated in Ion trap mode (operated in CI with isobutane, at 240°C).	0.1 - 0.5 ng/mL	2 – 5 ng/mL	(Cognard, Bouchonnet, and Staub 2006)

Compounds	Sample	Volume/weight (mL or mg)	Extraction procedure	GC conditions	LOD	LOQ	Ref.
Cocaine and benzoylecgonine	Oral fluid	0.25	SPE (UCT Clean Screen DAU (200 mg; 100 mL)). Derivatization with MTBSTFA + 1% tert-butyldimethylsilyl chloride (50:50 v/v)	2D-GC-MS: Capillary column (DBS-1MS,15 m × 0.25 mm, 0.25 μm and (ZB-50, 30 m × 0.32 mm, 0.25 μm). The initial oven temperature was held at 150°C for 0.5 min, followed by an increase of 30°C per minute until 290°C for 0.5 min (ramp 1), followed by an increase of 75°C per minute until 180°C for 1.5 min (ramp 2), the finally raising 30°C per minute until 290°C for 1 min (ramp 3). Post ramp: 40°C/min until 320°C for 4 min. The mass spectrometer was operated in EI mode	0.5 - 1 ng/mL	1 ng/mL	(Ellefsen et al. 2016)
cocaine, benzoylecgonine, ecgonine methyl ester, cocaethylene	Amniotic fluid and maternal serum	5000-30000	SPE (Chem-Elut column). The compounds were derivatized to n-propyl COC and p-fluoro-COC	GC/MS Capillary column (15m × 0.25-mm I.D., 0.10-μm film thickness) with 5% phenylmethylsiloxane (HP-5MS). The initial oven temperature was held at 100°C for 1 min, the followed by an increase of 30°C per minute until 280°C, maintained for 3 minutes. The temperatures of the injection port and the transfer line were set at 250°C and 280°C, respectively.	5 ng/mL	10 ng/mL	(Ripple et al. 1992)
Cocaine, cocaethylene, benzoylecgonine and anhydroecgonine methyl ester	Vitreous humor (*post-mortem*)	1000	Centrifugation (during 5 min at 1048 × g), SPE (Trace B® 335). Derivatization with MSTFA	GC-MS Capillary column (30m × 0.25-mm I.D., 0.25-μm film thickness) with 5% phenylmethylsiloxane (HP-5MS). The initial oven temperature was held at 90°C for 2 min, followed by an increase of 10°C per minute to 220°C, then increased 20°C per minute to 290°C, maintained for 4 min. The temperatures of the injection port and the transfer line were set at 220°C and 280°C, respectively. The mass spectrometer was operated with an electron energy of 70 eV in the positive EI mode. The temperature of the MS interface, the source and the quadrupole were 280°C, 230°C and 150°C, respectively.	1.0 – 2.0 ng/mL	10 ng/mL	(Peres et al. 2014)

Table 2. (Continued)

Compounds	Sample	Volume/weight (mL or mg)	Extraction procedure	GC conditions	LOD	LOQ	Ref.
Cocaine and benzoylecgonine	Breast milk	0.5	magnetic carbon nanotubes (agitation and solvent addition). Derivatization with BSTFA+1%TMCS	GC/MS Capillary column (30m × 0.25-mm I.D., 0.25-μm film thickness) with 5% diphenyl-95%dimethylpolisiloxane. The initial oven temperature was held at 150°C for 1 min, followed by an increase of 30°C per minute until 270°C, then finally raised at a rate of 30°C per minute until 280°C, maintained for 5 min. The temperatures of the injection port and the transfer line were set at 200°C and 300°C, respectively. The mass spectrometer was operated with an electron energy of 70 eV in the EI mode.	1.5 – 1.6 ng/mL	5.0 ng/mL	(Dos Santos et al. 2017)

Legend: 2D-GC-MS: two-dimensional-gas chromatography-mass spectrometry; BSTFA: N,O-bis-(trimethylsilyl)-trifluoroacetamide; DBS: Dried Blood Spots; EI: Electron Ionization; GC-MS: Gas chromatography coupled mass spectrometry; GC-MS/MS: Gas chromatography coupled mass spectrometry in tandem; HFIP: Hexafluoroisopropanol; HLB: Hydrophilic lipophilic balanced; HS-SPME: Head-space Solid Phase Microextraction; IT: Ion trap; LC-HRMS: Liquid Chromatography coupled to a High Resolution Mass Spectrometry; LLE: Liquid-Liquid Extraction; LOD: Limit of detection; LOQ: Limit of quantitation; LPME: Liquid Phase Microextraction; mCNTs: magnetic carbon nanotubes; MEPS: microextraction by packed sorbent; MSTFA: N-Methyl-N-trimethylsilyl-trifluoroacetamide; n.a.: not available; MTBSTFA: N-tert-Butyldimethylsilyl-N-methyltrifluoroacetamide; PCI: Positive Chemical Ionization PFPA: pentafluoropropionic anhydride; PFPOH: pentafluoropropanol; SIM: selected ion monitoring; SPE: solid-phase extraction; SPME: Solid-Phase Microextraction; TMCS: trimethylchlorosilane.

In the absence of the LOQ value, the lowest point of the calibration curve was considered.

Table 3. Review of the GC methods developed for cannabinoids determination in biological specimens

Compounds	Sample	Volume/weight (mL or mg)	Extraction procedure	GC conditions	LOD (ng/mL)	LOQ (ng/mL)	Ref.
THCCOOH	Human placenta	1000	SPE (STRATA™-X-C); Derivatization with MSTFA	GC-MS Capillary column: 5%-phenyl 95%-dimethylpolysiloxane (ZB-5, 15 m x 0.25 mm i.d., 0.25 mm); Oven temperature: 70°C (2 min), 30°C/min rate to 160°C, 5°C/min rate to 170°C, 20°C/min rate to 200°C, 10°C/min rate to 220°C, 30°C/min rate to 300°C (3 min)	1.3	3.9	(Joya et al. 2010)
THC, 11-OH-THC, CBD, CBN and THCCOOH	Oral fluid	1	PPT (ice-cold acetonitrile), SPE (CEREX® Polycrom™); Derivatization with TFAA and HFIP	2D-GC-MS (EI and NCI) Primary column: ZB-50 (30m × 0.25mm i.d., 0.25μm film thickness), secondary column: DB-1MS (15m × 0.25mm i.d., 0.25μm film thickness) (EI) and primary column: DB-1MS (15m × 0.25mm i.d., 0.25μm film thickness) and secondary column ZB-50 (30m × 0.32mm i.d., 0.25μm) (NCI);	0.5 (CBD), 0.5 (THC), 1 (CBN), 0.4 (11-OH-THC) and 0.006 (THCCOOH)	0.5 (CBD), 0.5 (THC), 1 (CBN), 0.5 (11-OH-THC) and 0.0075 (THCCOOH)	(Milman et al. 2010)

Table 3. (Continued)

Compounds	Sample	Volume/weight (mL or mg)	Extraction procedure	GC conditions	LOD (ng/mL)	LOQ (ng/mL)	Ref.
CBD, THC, THCCOOH and 11-OH-THC	Plasma	1	SPE (Styre Screen® SSTHC06Z); Derivatization with BSTFA and 1% TMCS	2D-GC-MS Primary column: ZB-50 (30 m × 0.25 mm × 0.25 µm), secondary column: DB-1MS (15 m × 0.25 mm × 0.25 µm); Oven temperature was held at 185°C initially for 0.5 min and increased at 45°C/min to 225°C and held for 3.0 min. A second ramp, 15°C/min to 275°C, was held at 275°C for 1.58 min. Secondary column: Oven temperature decreased at 80°C/min to 195°C and held for 2.3 min. During this hold time, cryotrap temperature increased at 700°C/min to 225°C. Oven temperature was ramped slowly 10°C/min to 230°C and then immediately 25°C/min to 275°C. EI mode	0.25 (CBD), 0.25 (THC), 0.125 (THCCOOH), 0.125 (11-OH-THC)	0.25 (CBD), 0.25 (THC), 0.125 (THCCOOH), 0.25 (11-OH-THC)	(Karschner et al. 2010)
CBD, CBN and THC	Hair	10	Alkaline hydrolysis (NaOH 1M), HF-LPME (Accurel® Q3/2 polypropylene) Addition of butyl acetate	GC-MS/MS Capillary column Varian VF-5MS (30m × 0.25mm i.d., 0.25 µm); Oven temperature: 60°C, 35°C/min	0.005 ng/mg (CBD), 0.005 ng/mg	0.001 ng/mg (CBD), 0.001 ng/mg (CBN),	(Emidio et al. 2010)

Compounds	Sample	Volume/weight (mL or mg)	Extraction procedure	GC conditions	LOD (ng/mL)	LOQ (ng/mL)	Ref.
				rate to 255°C (1min), 2°C/min rate to 280°C (2 min)	(CBN), 0.015 ng/mg (THC)	0.02 ng/mg (THC)	
CBD, CBN and THC	Hair	10	Alkaline hydrolysis (NaOH 1M), HS-SPME (PDMS fiber 100 μm). Addition of butyl acetate	GC-MS/MS (EI); Capillary column VF-5MS (30m×0.25mm i.d., 0.25 μm); Oven temperature: 100°C, 15°C/min rate to 280°C(6min)	0.007 ng/mg (CBD), 0.011 ng/mg (CBN), 0.031 ng/mg (THC)	0.012 ng/mg (CBD), 0.03 ng/mg (CBN), 0.062 ng/mg (THC)	(Emidio, Prata, and Dórea 2010)
THCCOOH	Hair	25	Alkaline hydrolysis (NaOH 1M), LLE (n-hexane:ethyl acetate (9:1); Derivatization with PFPOH and PFPA	GC-MS/MS (NCI); Capillary column (DB-5MS, 30m x 0.25 mm i.d., 0.25 mm); 100°C (1 min), 35°C/min rate to 275°C (3 min), 25°C/min rate to 300°C (3.5 min)	0.000015 ng/mg	0.00005 ng/mg	(Kim et al. 2011)
THC and THCCOOH	Hair	50	Alkaline hydrolysis (NaOH 1M), LLE (n-hexane:ethyl acetate (9:1); Derivatization with PFPA and HFIP	GC-MS/MS (NCI) (THCCOOH) and GC-MS/MS (EI) (THC) Capillary column Varian CP-Sil 8 (15 m x 0.25 mm i.d.,0.25 mm); 70°C, 15°C/min rate to 200°C, 60°C/min rate to 320°C (THC) and 100°C, 25°C/min rate to210°C, 10°C /min rate to 240°C, 60°C/min rate to 300°C (THCCOOH)	0.00001 ng/mg	0.00004 ng/mg	(Minoli et al. 2012)

Table 3. (Continued)

Compounds	Sample	Volume/weight (mL or mg)	Extraction procedure	GC conditions	LOD (ng/mL)	LOQ (ng/mL)	Ref.
THCCOOH	Urine	3	Hydrolysis (KOH 10M) and SPE (Bond Elut Certify II, C8); Derivatization with BSTFA	GC-MS/MS Capillary column (HP-5MS, 30m × 0.25 mm i.d., 0.25 μm); 100°C (1 min), 35°C/min rate to 200°C, 25°C/min rate to 250°C, 20°C/min rate to 310°C (9 min)	0.5	1	(V. de M. Prata, Emídio, and Dorea 2012)
CBD, CBN, THC, 11-OH-THC and THCCOOH	Blood (postmortem)	1	LLE (hexane-ethyl acetate (5:1); Derivatization with MSTFA	2D-GC-MS (EI) Primary column: HP-5MS (30 m x 0.25 mm i.d., 0.25 mm), secondary column:DB-17MS (15 m x 0.32 mm i.d.,0.25 mm). Oven temperature: 150°C (1 min), 50°C/min rate to 220°C, 30°C/min rate to 280°C, 120°C/min rate to 185°C and 10°C/min to 280°C. Inlet temperature 250°C. MS was operated in EI mode	0.25	0.25	(Andrews and Paterson 2012)
THCA	Urine	1	Hydrolysis (NaOH 6M) and LLE (n-hexane/ethyl acetate (9:1)); Derivatization with MSTFA and MSTFA/ethanethiol/NH4I (500:4:2)	GC-MS/MS Capillary column: HP-1MS (12 m x 250 mm, 0.25 mm); 110°C (0.15 min), 70°C/min rate to 310°C (1 min)	0.0567	0.1889	(De Brabanter et al. 2013)

Compounds	Sample	Volume/weight (mL or mg)	Extraction procedure	GC conditions	LOD (ng/mL)	LOQ (ng/mL)	Ref.
THC and THCCOOH	Vitreous humor	1	SPE (Trace Bw 335); Derivatization with MSTFA	GC-MS (EI) Capillary column: HP-5MS fused silica (30 m x 0.25 mm i.d., 0.25 mm); 90°C (2 min), 10°C/min rate to 220°C, 20°C/min rate to 290°C (4 min)	2	--	(Peres et al. 2014)
THCCOOH	Oral fluid	0.250	PPT (ice-cold acetonitrile) and SPE (CEREX Polychrom); Derivatization with TFAA and HFIP	GC-MS/MS Capillary column: HP 5MS (15 m × 0.25 mm, 0.25-μm); 175°C (1 min), 30°C/min rate to 275°C (0.9 min)	0.0075	0.01	(Barnes, Scheidweiler, and Huestis 2014)
THC and THCCOOH	Blood (post-mortem)	1	SPE (Bond Elut Certify, C8); Derivatization with MSTFA	GC-MS Capillary column: HP-5MS (30 m x 0.25 mm i.d., 0.25-mm); 90°C (2 min), 10°C/min rate to 220°C, 30°C/min rate to 290°C (6 min)	5	10	(Pelição et al. 2014)
THC, 11-OH-THC and THCCOOH	Plasma and Serum	1	SPE (Chromabond®, C18); Derivatization with MSTFA	GC-MS Capillary column: OPTIMA® 5 MS Accent 95% dimethylpolysiloxane, 5% diphenylpolysiloxane, (30 m× 0.25 mm i.d.× 0.25); 150°C (1.5min), 9°C/min rate to 260°C (6 min), 30°/min rate to 300°C (8 min)	Plasma: 0.15 (THC), 0.25 (11-OH-THC), 2 (THCCOOH); Serum: 0.15 (THC), 0.25 (11-OH-THC), 1.6 (THCCOOH)	Plasma: 0.3 (THC), 0.35 (11-OH-THC), 3 (THCCOOH); Serum: 0.3 (THC), 0.35 (11-OH-THC), 3.3 (THCCOOH)	(Gasse et al. 2016)

Table 3. (Continued)

Compounds	Sample	Volume/weight (mL or mg)	Extraction procedure	GC conditions	LOD (ng/mL)	LOQ (ng/mL)	Ref.
THC, 11-OH-THC and THCCOOH	Serum	1 (manual LLE) and 0.5 (automated LLE)	Manual LLE (n-hexane-ethyl acetate (9:1) and automated LLE (n-hexane-ethyl acetate (9:1); Derivatization with MSTFA (manual LLE) and MSTFA/ethyl acetate (3/2, v/v) (automated LLE)	GC-MS Capillary column: Optima 5 HT (30 m × 0.25 mm, 0.25 μm) (manual LLE) and DB-5-MS (30 m × 0.25 mm, 0.25 μm) (automated LLE); 70°C (2 min), 20°C/min rate to 250°C (4 min), 20°C/min rate to 300°C (17 min) (manual LLE) and 70°C (2 min), 20°C/min rate to 240°C (5 min), 20°C/min rate to 250°C (6 min), 20°C/min rate to 260°C (5 min)	0.3 (THC), 0.1 (11-OH-THC) and 0.3 (THCCOOH)	0.6 (THC), 0.8 (11-OH-THC) and 1.1 (THCCOOH)	(Purschke et al. 2016)
THC, CBN and CBD	Hair	120–150	Alkaline hydrolysis (NaOH 1M), LLE (n-hexane-ethyl acetate (9:1); Derivatization with BSTFA and TMCS	GC-MS Capillary column: DB-5MS (30 m × 0.25 mm, 0.25 μm)	0.003 ng/mg (THC), 0.01 ng/mg (CBN) and 0.01 ng/mg (CBD)	0.01 ng/mg (THC), 0.06 ng/mg (CBN) and 0.03 ng/mg (CBD)	(Heinl, Lerch, and Erdmann 2016)
THC, 11-OH-THC and THCCOOH	Plasma	0.25	PPT (frozen acetonitrile (3:1)) and MEPS (4 mg; 80% C8 and 20% SCX); Derivatization with MSTFA and 5% TMCS	GC-MS/MS Capillary column: HP-5MS with 5% phenylmethylsiloxane (30 m × 0.25-mm I.D., 0.25 μm); 120°C (1.85 min), 20°C/min rate to 300°C (9.10 min)	0.1	0.1	(T Rosado et al. 2017)

Compounds	Sample	Volume/weight (mL or mg)	Extraction procedure	GC conditions	LOD (ng/mL)	LOQ (ng/mL)	Ref.
THC, CBD and CBN	Teeth	500	SPE (HCX); Derivatization with BSTFA and 1% TMCS	GC-MS Capillary column: ZB-5MS (20 m, 0.18mm; 0.18 μm); 100°C (1 min), 40°C/min rate to 220°C (1 min), 10°C/min rate to 310°C	0.01 (THC), 0.04 (CBD) and 0.03 (CBN)	0.05	(Ottaviani et al. 2017)
8β-OH-THC and 8β,11-diOH-THC	Plasma	1	SPE (Chromabond® C8); Derivatization with MSTFA	GC-MS Capillary column: OPTIMA® 35 MS Accent (65% dimethylpolysiloxane, 35% diphenylpolysiloxane, 30 m × 0.25 mm i.d. × 0.25 μm); 150°C (1.5 min), 10°C/min rate to 260°C (9 min), 30°C/min rate to 300°C (10 min)	0.2 (8β-OH-THC) and 0.3 (8β,11-diOH-THC)	0.3 (8β-OH-THC) and 0.4 (8β,11-diOH-THC)	(Gasse et al. 2018)
THC, 11-OH-THC and THCCOOH	Whole blood and urine	1	Hydrolysis (NaOH 1M), LLE (acetonitrile, triethylamine, propylchloroformate and hexane) (urine) and deproteinization (acetonitrile), LLE (NaOH, triethylamine, propylchloroformate and hexane) (whole blood); Derivatization with PCF	GC-MS Capillary column: Rtx® -5MS Crossbond® 5% diphenyl–95% dimethylpolysiloxane (15m x 0.25mm i.d. x 0.25 μm); 100°C (60 s), 17°C/min rate to 320°C (1 min)	Urine: 0.5 (THC) and 1.2 (THCCOOH); whole blood: 0.2 (THC), 0.2 (OH-THC) and 0.9 (THCCOOH)	Urine: 1.3 (THC) and 2.6 (THCCOOH); whole blood: 0.5 (THC), 0.6 (OH-THC) and 2.4 (THCCOOH)	(Stefanelli et al. 2018)

Table 3. (Continued)

Compounds	Sample	Volume/weight (mL or mg)	Extraction procedure	GC conditions	LOD (ng/mL)	LOQ (ng/mL)	Ref.
THCCOOH	Meconium	500	Hydrolysis (NaOH 4M) and SPE (Bond Elut II (C8)); Derivatization with N-tert-butyldimethylsilyl-I-N-methyltrifluoroacetamide with 1% TBSTFA	GC-MS Capillary column: HP-5MS (30m × 0.25mm × 0.1 μm); 120°C (2 min), 20°C/min rate to 290°C (2 min)	5 ng/g	10 ng/g	(Mantovani et al. 2018)
THC, CBD, CBN and 11-OH-THC	Hair	50	Hydrolysis (NaOH 1M) and LLE (n-hexane-ethyl acetate (9:1)); Derivatization with MSTFA	GC-MS/MS (EI); column: DB5-MS (15 m × 0.25 mm i.d., 0.25 μm); 100°C (1 min), 20°C/min rate to 280°C (2 min), 40°C/min rate to 320°C (1 min) (THC,CBD, CBN, 11-OH-THC)	600 ng/mg (THC), 1400 ng/mg (CBD), 300 ng/mg (CBN), 30 (11-OH-THC)	1900 ng/mg (THC), 4700 ng/mg (CBD), 900 ng/mg (CBN), 100 ng/mg (11-OH-THC)	(Angeli et al. 2018)
CBD, CBN, THC, 11-OH-THC and THCCOOH	Hair	25	Hydrolysis (NaOH 1M) and SPE (CHROMABOND HR-XA); Derivatization with MSTFA	GC-MS/MS Capillary column: Zebron ZB-5MSi (30 m x 0.25 mm x 0.25 μm); 100°C (1 min), 15°C/min rate to 205°C, 10°C/min rate to 310°C	0.002 ng/mg (CBD), 0.002 ng/ml (CBN), 0.002 ng/mg (THC), 0.0002 ng/mg (11-OH-THC) and 0.0002 ng/mg (THCCOOH)	0.005 ng/mg (CBD), 0.005 ng/ml (CBN), 0.005 ng/mg (THC), 0.0005 ng/mg (11-OH-THC) and 0.0005 ng/mg (THCCOOH)	(Kieliba et al. 2019)

Legend: 2D-GC-MS: two-dimensional gas chromatography mass spectrometry; 8β-OH-THC: 8β-hydroxy-Δ9-tetrahydrocannabinol; 8β,11-diOH-THC: 8β,11-dihydroxy-Δ9-tetrahydrocannabinol; 11-OH-THC: 11-hydroxy- tetrahydrocannabinol; BSTFA: N,O-bis-(trimethylsilyl)-trifluoroacetamide; CBD: cannabidiol; CBN: cannabinol; EI: Electron Ionization; HFIP: Hexafluoroisopropanol; GC-MS: Gas chromatography coupled mass spectrometry; GC-MS/MS:

Gas chromatography coupled mass spectrometry in tandem; HF-LPME: Hollow fiber-based liquid phase microextraction; HS-SPME: headspace solid-phase microextraction; LLE: Liquid-liquid extraction; LOD: Limit of detection; LOQ: Limit of quantitation; MEPS: microextraction by packed sorbent; NCI: Negative Chemical Ionization; MSTFA: N-Methyl-N-trimethylsilyl-trifluoroacetamide; MTBSTFA: N-tert-Butyldimethylsilyl-N-methyltrifluoroacetamide; PCF: propyl chloroformate; PFPA: pentafluoropropionic anhydride; PFPOH: pentafluoropropanol; PPT: Protein precipitation; SPE: Solid-phase extraction; SPME: solid-phase microextraction; TFAA: trifluoroacetic anhydride; THC: Δ9-tetrahydrocannabinol; THCA: 11-nor-D9-tetrahydrocannabinol-9-carboxylic acid; THCCOOH: 11-nor-9-carboxy-tetrahydrocannabinol; TMCS: trimethylchlorosilane.

In the absence of the LOQ value, the lowest point of the calibration curve was considered.

Alvear et al. (Alvear et al. 2014) conducted a multi-sample study in order to determine cocaine and benzoylecgonine using SPE and GC-MS. As the samples were *postmortem* and were of either solid (liver, accumbens nucleus and ventral tegmental area) or liquid (venous and arterial blood, femoral arterial and venous blood, urine, vitreous humor and cerebrospinal fluid) nature, it was necessary to foil the differences between these two types of samples, and it was verified that the filtration step was slightly different, which were surpassed throughout the validation procedure. The LODs obtained varied from 0.19 to 5.09 ng/mL or g (cocaine) and 0.76 to 6.00 ng/mL or g (benzoylecgonine) and the LOQs from 0.62 to 16.95 ng/mL or g (cocaine) and 2.55 to 20.00 ng/mg or g (benzoylecgonine). With such results it was possible to verify that the greatest quantity of analytes was found in urine. Overall, the method proved to be useful by choosing different matrices than the usual blood. Moreover, it serves as a tool to further cases of deaths of drug users.

Dos Santos et al. (Dos Santos et al. 2017) developed a new method to detect cocaine and benzoylecgonine in breast milk samples using a new extraction technique: magnetic carbon nanotubes (mCNTs) as a replacement adsorbent for SPE and SPME techniques, in order to minimize the use of organic solvents using a simple and miniaturized technique. The method was fully optimized and allowed to accomplish good linearity, selectivity and sensitivity, with LODs in the range of ng/mL. Therefore, it was considered a suitable method for further routine analyses. Table 2 resumes analytical protocols to determine cocaine and metabolites in biological specimens.

DETERMINATION OF CANNABINOIDS

Cannabis has been one of the most widely used drugs of abuse worldwide, and also one of the most commonly detected in clinical and forensic laboratories (Stefanelli et al. 2018; Gasse et al. 2018). The main psychoactive constituent of cannabis is Δ9-tetrahydrocannabinol (THC), and other compounds such as cannabidiol (CBD) and cannabinol (CBN) are also present (Stefanelli et al. 2018; Gasse et al. 2018). After entering the body, THC is oxidized by cytochrome P450 CYP2C9 and CYP2C19 to 11-hydroxy-Δ9-tetrahydrocannabinol (11-OH-THC), which is in turn metabolised to 11-nor-9-carboxy-9-tetrahydrocannabinol (THC-COOH) (Gasse et al. 2018). These phase I metabolites also undergo phase II reactions and are metabolised by UDP-glucuronosyltransferases (Gasse et al. 2018). These types of compounds have a highly lipophilic character, accumulating in adipose tissues and being released later (Rosado et al. 2017). 8β-hydroxy-Δ9-tetrahydrocannabinol (8β-OH-THC) and 8β,11-dihydroxy-Δ9-tetrahydrocannabinol (8β, 11-diOH-THC) metabolites are described as well (Gasse et al. 2018).

Given the increased consumption of this illicit substance, various methods of detection and quantification of cannabis have been developed in various plant and biological samples. Both LC and GC-based methods have been used for such purposes. Béres et al. (Béres et al. 2019) recently conducted a study comparing phytocannabinoid analysis using both chromatographic techniques. It was found that GC showed better results, namely in the detection of the less stable compounds, such as phytocannabinoid acids (Béres et al. 2019). Additionally, GC methods for cannabinoid quantification are usually faster and simpler than some types of LC, namely high performance liquid chromatography (HPLC) (Nahar, Guo, and Sarker 2019).

GC has been shown to be an important tool for plant material analysis. This type of chromatographic technique allows the quantification of CBD, CBN, THC and metabolites, and can be used to define the plant phenotype and to determine the purity of the drug or even from which plant constituent it is coming (Raharjo and Verpoorte 2004). The GC technique can also be used to detect cannabinoids in wastewaters (Racamonde et al. 2012).

GC has been widely applied for biological samples analysis in clinical and forensic context, and some methods developed in the last ten years are compiled in Table 3. Blood is one of the most used samples for analytical determinations (Stefanelli et al. 2018; Pelição et al. 2014; Andrews and Paterson 2012; Rosado et al. 2018). Although this sample has the disadvantage of being collected invasively, it presents the possibility of detecting the unchanged parent drug (Rosado et al. 2018). Similarly, plasma and serum are other samples widely used for the detection of cannabinoids use, although they present the same disadvantages (Gasse et al. 2018; Rosado et al. 2017; Purschke et al. 2016; Gasse et al. 2016; Karschner et al. 2010). Rosado et al. (Rosado et al. 2017) developed a sensitive analytical method in which they quantified THC, 11-OH-THC and THCCOOH in a GC-MS/MS instrument using only 0.25 mL of plasma, showing a LOD and LOQ of 0.1 ng/mL. Urine is another biological matrix that has been used for cannabinoid assay (Stefanelli et al. 2018; De Brabanter et al. 2013; V. de M. Prata, Emídio, and Dorea 2012; Nahar, Guo, and Sarker 2019). This matrix has the advantage of presenting a wider detection window, allowing compounds detection up to 12 days after consumption (Raharjo and Verpoorte 2004). The major compound excreted in this matrix includes both conjugated and unconjugated forms of THC-COOH (Raharjo and Verpoorte 2004). Stefanelli et al. (Stefanelli et al. 2018) developed an analytical method by GC-MS where they used 1 mL of urine and blood and were able to quantify THC, 11-OH-THC and THCCOOH. In recent years, alternative biological samples have been used to determine this class of compounds. Oral fluid is one of the alternative matrices that has been used for cannabinoid detection (Barnes, Scheidweiler, and Huestis 2014; Milman et al. 2010). This biological specimen allows to correlate the amounts detected in it with the levels of compounds present in blood (Gonçalves et al. 2019). The use of oral fluid has also the advantage of a non-invasive collection procedure, but the quantification of cannabinoids in this matrix may be influenced by the presence of drugs and food (Gonçalves et al. 2019). Barnes et al. (Barnes, Scheidweiler, and Huestis 2014) developed an analytical method for THCCOOH quantification using GC-MS/MS equipment and using only 0.25 mL of oral fluid, obtaining a LOD of 0.0075 ng/mL and a LOQ of 0.01 ng/mL. Another sample that has been shown to be of great interest in clinical and forensic analysis is hair (Kieliba et al. 2019; Angeli et al. 2018; Heinl, Lerch, and Erdmann 2016; Minoli et al. 2012; Kim et al. 2011; Emídio, Prata, and Dórea 2010; Emídio et al. 2010). This biological sample has a wider detection window (from weeks to months or even years), is less susceptible to changes caused by metabolism and is more difficult to tamper with (Gonçalves et al. 2019; Rosado et al. 2018). Several methods for the detection and quantification of cannabinoids have been developed in this biological matrix (Kieliba et al. 2019; Angeli et al. 2018; Heinl, Lerch, and Erdmann 2016; Minoli et al. 2012; Kim et al. 2011; Emídio, Prata, and Dórea 2010; Emídio et al. 2010). Emídio et al. (Emídio et al. 2010) developed a method for the determination of cannabinoids by GC-MS/MS using only 10 mg of hair. The method presented an LOD between 0.005 ng/mg and 0.015 ng/mg (Emídio et al. 2010). However, there are some things to bear in mind when using hair to avoid results misinterpretation. For instance, cosmetically treated hair (e.g., darker treatments) may accumulate basic compounds more easily (Heinl, Lerch, and Erdmann 2016). Additionally, this matrix presents a high risk of external contamination (Gonçalves et al. 2019). Therefore, detection of THC-COOH is mandatory since this compound is formed *in vivo* after cannabis use, excluding the possibility of external contamination (Kieliba et al. 2019; Angeli et al. 2018). However, only more sophisticated equipment allows the determination of this particular compound in this particular specimen, namely GC-MS with negative chemical ionization (NCI), GC-MS/MS and 2D-GC (Kieliba et al. 2019; Angeli et al. 2018; Pragst 2004; Gonçalves et al. 2019). The use of GC-MS with NCI is considered the preferred method for cannabinoids detection in hair, but NCI has a reduced lack of flexibility compared to electronic ionization (EI) (Kieliba et al. 2019; Nahar, Guo, and Sarker 2019). Other unconventional samples have been used to develop analytical methodologies to quantify cannabinoids, namely teeth (Ottaviani et al. 2017), vitreous humor (Peres et al. 2014), human placenta (Joya et al. 2010) and sweat (Gentili et al. 2016).

Cannabinoid use during pregnancy has also been studied. It is estimated that about 10% of pregnant women in Europe are cannabinoid users, while in Brazil and the United States these figures are 4.1% and 3.4%, respectively (Mantovani et al. 2018). Thus, biological matrices such as meconium have been used to develop and validate analytical methods to monitor cannabinoid use during pregnancy (Mantovani et al. 2018). Mantovani et al. (Mantovani et al. 2018) developed an analytical method for the quantification of cannabinoids in meconium samples. For this, they used a GC-MS equipment, obtaining a LOD of 5 ng/mg and a LOQ of 10 ng/mg.

When analysing biological samples for cannabinoids detection and quantification by GC it is important to include a sample preparation step (Raharjo and Verpoorte 2004). Sample pre-treatment should be simple and reproducible, allowing for increased sensitivity and a reduction in the number of interferences (Raharjo and Verpoorte 2004). For cannabinoids analysis, classical techniques such as liquid-liquid extraction (LLE) and SPE continue to be widely used (Kieliba et al. 2019; Angeli et al. 2018; Mantovani et al. 2018; Stefanelli et al. 2018; Gasse et al. 2018; Ottaviani et al. 2017; Heinl, Lerch, and Erdmann 2016; Purschke et al. 2016; Gasse et al. 2016; Pelição et al. 2014; Barnes, Scheidweiler, and Huestis 2014; Peres et al. 2014; De Brabanter et al. 2013; Andrews and Paterson 2012; V. de M. Prata, Emídio, and Dorea 2012; Minoli et al. 2012; Kim et al. 2011; Karschner et al. 2010; Milman et al. 2010; Joya et al. 2010), being n-hexane:ethyl acetate (9:1) the most common solvent used (Angeli et al. 2018; Heinl, Lerch, and Erdmann 2016; Purschke et al. 2016; De Brabanter et al. 2013; Andrews and Paterson 2012; Minoli et al. 2012; Kim et al. 2011), as it usually provides recovery values greater than 90% (Raharjo and Verpoorte 2004).

A greater variety exists regarding SPE, but it is quite common to use reverse phase columns (Raharjo and Verpoorte 2004). In recent years miniaturized extraction techniques, such as hollow fiber-based liquid phase microextraction (HF-LPME) (Emídio et al. 2010), headspace solid-phase microextraction (HS-SPME), SPME (Emídio, Prata, and Dórea 2010) and MEPS (Rosado et al. 2017) have been implemented. These techniques allow a reduction in preparation times and sample volumes, as well as the amount of extraction solvents used (Gonçalves et al. 2019). Additionally, miniaturized techniques have advantages such as good extraction efficiencies, reduced cost of operation and the ability to engage online with analytical instruments (Gonçalves et al. 2019). Associated to extraction techniques, it is also sometimes necessary to include protein precipitation (PPT), hydrolysis or washing steps (Raharjo and Verpoorte 2004).

Currently, several stationary phases are available for cannabinoids detection by GC. Generally, fused silica chromatographic columns are used, namely ZB-5, ZB-50, DB-5 MS, HP-5 MS, also having a small diameter (Angeli et al. 2018; Mantovani et al. 2018; Ottaviani et al. 2017; Rosado et al. 2017; Heinl, Lerch, and Erdmann 2016; Pelição et al. 2014; Barnes, Scheidweiler, and Huestis 2014; Peres et al. 2014; Kim et al. 2011; Karschner et al. 2010; Milman et al. 2010; Joya et al. 2010; Nahar, Guo, and Sarker 2019; Raharjo and Verpoorte 2004). Selecting the column is an important step in ensuring that the analysis is performed successfully, however in the particular case of cannabinoids almost all commercially available columns allow proper separation of these compounds from biological samples (Nahar, Guo, and Sarker 2019; Raharjo and Verpoorte 2004). Regarding cannabinoid detection when using GC, two detectors can be distinguished: FID and MS (Nahar, Guo, and Sarker 2019). GC-FID is a reliable and simple device, so it is mainly used in routine analysis (Nahar, Guo, and Sarker 2019). However, GC coupled to a MS detector is considered the preferred technique for the identification of cannabinoids at the forensic and clinical level, because it allows identifying unequivocally the compounds (Nahar, Guo, and Sarker 2019). In recent years, it has been possible to create compound libraries, given the quality of information gathered about cannabinoid structures.

Thus, MS has also been improving and the development of analytical methods using MS-MS has emerged (Kieliba et al. 2019; Angeli et al. 2018; Rosado et al. 2017; Barnes, Scheidweiler, and Huestis 2014; De Brabanter et al. 2013; Prata, Emídio, and Dorea 2012; Minoli et al. 2012; Kim et al. 2011; Emídio, Prata, and Dórea 2010; Emídio et al. 2010). This form of detection is often applied not only to biological samples but also to natural matrices (Nahar, Guo, and Sarker 2019). EI is the most frequent mode of analysis when GC-MS is used, as it allows obtaining ion fragments essential for cannabinoid identification (Nahar, Guo, and Sarker 2019). Additionally, EI improves LOD and increases sensitivity and specificity (Shah et al. 2019).

However, this type of ionization has a high extent of fragmentation and consequently it is not always possible to detect the molecular ion (Nahar, Guo, and Sarker 2019). Other less extensive ionization modes are also often used for cannabinoids analysis, namely NCI. This ionization mode is smoother and usually allows obtaining information about the molecular ion (Nahar, Guo, and Sarker 2019). With regard to cannabinoid analysis, photo ionization (PI) can be highlighted, but its use is less frequent (Nahar, Guo, and Sarker 2019). In recent years 2D-GC has emerged, which has the advantages of increasing sensitivity and improving peak shape (Shah et al. 2019). 2D-GC can be used in NCI mode and coupled to a MS detector, allowing separating structurally similar cannabinoids (Shah et al. 2019; Aizpurua-Olaizola et al. 2014).

Cannabinoids analysis is not always simple, and as such care should be taken when handling the sample, especially when analysing acidic cannabinoids (Raharjo and Verpoorte 2004). These compounds when exposed to high temperatures may undergo decarboxylation, becoming neutral (Raharjo and Verpoorte 2004). This process can occur when the sample is injected into the GC equipment, so only cannabinoids that do not have carboxyl groups can be injected directly into the GC equipment (Raharjo and Verpoorte 2004). Therefore, it is important to include a derivatization step to prevent the loss of the carboxyl group in acidic cannabinoids (Raharjo and Verpoorte 2004). With this process it is possible to increase the volatility and stability of the cannabinoids present in the samples, as well as to improve the ionization and detectability of the compounds, since high electron affinity groups are introduced into the analyte structure (Raharjo and Verpoorte 2004; Shah et al. 2019). For the analysis of cannabinoids in biological samples, the most commonly used derivatizing solvents are acylating agents such as TFAA, MBTFA, PFPA and PFPOH (Raharjo and Verpoorte 2004). Derivatization processes are also crucial for phytocannabinoid analysis, since they are non-volatile compounds (Nahar, Guo, and Sarker 2019). For plant samples, derivatization processes include trimethylsilylation reactions with MSTFA, BSTFA and MTBSTFA (Raharjo and Verpoorte 2004). Also, the use of agents promoting alkylation reactions, such as trimethylanilium hydroxide

(TMAH) and tetrabutylammonium hydroxide (TBH), are often used in cannabinoid analysis (Raharjo and Verpoorte 2004). However, the inclusion of a derivatization step has some disadvantages, namely a more complex and time-consuming sample pre-treatment procedure (Nahar, Guo, and Sarker 2019; Raharjo and Verpoorte 2004; Shah et al. 2019).

DETERMINATION OF AMPHETAMINES AND AMPHETAMINES DERIVATIVES

In the last years, there has been an intensification in the use of amphetamines and derivatives, consequently increasing the interest of their determination and analysis in both clinical and forensic toxicology. It should be noted that, in addition to these compounds being detected in several conventional or non-conventional biological samples, there are several analytical methods used for their identification using gas chromatography, which is emphasised in this chapter. However, there are several challenges in identifying these drugs in specimens, such as the large number of potential structures and the introduction of new compounds on the illicit market almost every day.

Table 4 represents the compilation of the selected articles of the methods developed for the determination of amphetamines and their derivatives in biological specimens.

Due to the stimulant and hallucinogenic effects of amphetamines and their derivatives, resulting in a significant public health problem, efforts are increasingly being made to develop methodologies capable of detecting and determining these illicit drugs. These compounds have been reported in both clinical and forensic cases and, in an attemp to combat their use as an abuse drug, it is important to detect them in low concentrations, using simple, rapid, selective and sensitive methodologies (Chalavi et al. 2019).

The most commonly used biological sample for the identification of amphetamines is urine, considered a conventional specimen due to the ubiquity of this sample in clinical cases and the ease of collecting and processing it.

An example of the use of this specimen is the work developed by Mercieca et al. (Mercieca et al. 2018) for the determination of these illicit compounds in samples of 2 mL of urine and whole blood. Using the dispersive liquid-liquid microextraction technique (DLLME) for sample preparation, they obtained for urine a LOQ of 50 ng/mL for amphetamine, 2 ng/mL for methamphetamine, 5 ng/mL for 4-methylamphetamine, 3,4-methylenedioxymethamphetamine (MDMA), 3,4-methylenedioxy-amphetamine (MDA) and 3,4-methylenedioxy-N-ethylamphetamine (MDE) and 10 ng/mL for 4-methylthioamphetamine. The authors concluded that the low cost, speed and use of GC-MS makes the developed method applicable in a wide range of forensic and clinical toxicology laboratories.

It should be noted that more recently, Malaca et al. (Malaca et al. 2019) developed a new method for the determination of amphetamine and five of its derivatives in only 200 µL of urine samples. Using MEPS, they obtained a LOQ of 25 ng/mL for methamphetamine, 3,4-methylenedioxy-N-methyl-α-ethylfenilethylamine and MDE, a value of 35 ng/mL for amphetamine and MDMA and a value of 50 ng/mL for MDA. The authors reported that the described method proved to be easy and fast to perform, with the possibility of reutilization of the sorbent, profitable by minimizing solvents and sample, being in addition sensitive, selective and precise, which makes it an alternative to the classic techniques.

In recent years, there has been further development of analytical instrumentation to determine analytes in complex matrices as an alternative to conventional samples, both in clinical and forensic cases. As these compounds are consumed illegally, it is essential that laboratories have several analytical resources, thinking that consumers can hide the use of these drugs.

Mohamed et al. (K. Mohamed 2017) have developed a method for the quantification of amphetamine, methamphetamine, 4-methylamphetamine, MDA, MDMA and MDE in 500 µL of oral fluid. They achieved a LOQ value of 2.5 ng/mL for all compounds and the method was applied to oral fluid samples collected from consumers.

Table 4. Review of the methods developed for amphetamines and amphetamine derivatives determination in biological specimens

Compounds	Sample	Volume/weight (mL or mg)	Extraction procedure	GC Conditions	LOD (ng/mL)	LOQ (ng/mL)	Ref.
Amphetamine, Methamphetamine, 3,4-methylenedioxyamphetamine, 3,4-Methylenedioxy-N-ethylamphetamine, 3,4-methylenedioxy-N-methyl-α-ethylfenilethylamine and 3,4-Methylenedioxymethamphetamine	Urine	0.2	MEPS (C18 cartridge) and Derivatization with MBTFA	GC-MS (EI) Capillary column 5% of phenylmethylsilcxane. The temperature of the oven started at 90°C for the initial 2 min, increasing 20°C per minute up to 300°C where it was maintained for 3 min. Inlet and transfer line temperatures were set at 250°C and 280°C, respectively.	n.a.	25 (Methamphetamine, 3,4-Methylenedioxy-N-ethylamphetamine, 3,4-methylenedioxy-N-methyl-α-ethylfenilethylamine); 35 (Amphetamine and 4-Methylenedioxymethamphetamine); 50 (3,4-methylenedioxyamphetamine)	(Malaca et al. 2019)
3,4-methylenedioxyamphetamine, 3,4-methylenedioxymethamphetamine and 3,4-methylenedioxyethylamphetamine	Vitreous Humor	1	LLE (ethyl acetate) and Derivatization with HFAA	GC-MS (EI) Capillary column DB-5MS (5% phenyl/95% dimethyl polysiloxane). Oven temperature: 70°C with a holding time of 2 min then increased at 15°C/min rate to 280°C (3min). The injector, trap and the transfer line were set to 250°C, 200°C and 230°C, respectively.	1 for all, except 2.5 for 3,4-methylenedioxyamphetamine	10	(Ishikawa et al. 2018)

Table 4. (Continued)

Compounds	Sample	Volume/weight (mL or mg)	Extraction procedure	GC Conditions	LOD (ng/mL)	LOQ (ng/mL)	Ref.
Amphetamine, Methamphetamine, 4-methylamphetamine, 4-methylthioamphetamine, 3,4-methylenedioxymethamphetamine, 3,4-methylenedioxyamphetamine and 3,4-methylenedioxy-N-ethylamphetamine	Blood Urine	2	DLLME (chloroform as extractant and methanol as disperser solvent) and Derivatization with hexyl chloroformate	GC-MS (EI) Capillary column 5% phenyl-methylsilicone Oven temperature was held at 130°C for 2 min, increased to 270°C at 15°C/min, and a final temperature ramp at 50°C/min, to 310°C (held for 4 min). The injection port was set at 270°C.	50 (Amphetamine, 3,4-methylenedioxyamphetamine); 5 (Methamphetamine, 4-methylthioamphetamine, 3,4-methylenedioxymethamphetamine); 10 (4-methylamphetamine); 2 (3,4-methylenedioxy-N-ethylamphetamine) for blood samples; 50 (Amphetamine); 2 (Methamphetamine, 4-methylamphetamine, 4-methyl-	50 (Amphetamine, 3,4-methylenedioxyamphetamine); 10 (Methamphetamine, 4-methylamphetamine, 4-methylthioamphetamine, 3,4-methylenedioxymethamphetamine); 2 (3,4-methylenedioxy-N-ethylamphetamine) for blood samples; 50 (Amphetamine); 2 (Methamphetamine); 5 (4-methylamphetamine, 3,4-methylenedioxymethamphetamine, 3,4-methylenedioxyamphetamine, 3,4-methylenedioxy-N-ethylamphetamine);10 (4-methylthioamphetamine) for urine samples	(Mercieca et al. 2018)

Compounds	Sample	Volume/weight (mL or mg)	Extraction procedure	GC Conditions	LOD (ng/mL)	LOQ (ng/mL)	Ref.
					thioamphetamine, 3,4-methylenedioxymethamphetamine, 3,4-methylenedioxyamphetamine, 3,4-methylenedioxy-N-ethylamphetamine) for urine samples		
Amphetamine, Methamphetamine	Maternal and Neonate Meconium Hair Urine	400 for meconium, 24-60 for hair and 1 for urine	SPE (Bond Elut Certify) and Derivatization with ethylacetate and HFAA	GC-MS (EI) Capillary column DB-5. The column temperature was kept for 0.8 min at 150°C, increased to 210°C at 20°C/min for 5 min and afterwards incremented to 300°C at 25°C/min for 2 min. The temperature of the injection port was 280°C	50 ng/g for meconium samples; 1 ng/mg for hair samples; 50 ng/mL for urine samples	100 ng/g for meconium samples; 2 ng/mg for hair samples; 100 ng/mL for urine samples	(Jarque et al. 2018)

Table 4. (Continued)

Compounds	Sample	Volume/weight (mL or mg)	Extraction procedure	GC Conditions	LOD (ng/mL)	LOQ (ng/mL)	Ref.
Amphetamine, Methamphetamine, 4-methyl-amphetamine, 3,4-methylenedioxyamphetamine, 3,4-methylenedioxymethamphetamine and 3,4-methylenedioxyethylamphetamine	Oral Fluid	0.5	LLE (Addition of saturated $KHCO_3$ and ethyl acetate). Derivatization with ethyl chloroformate	GC-MS (EI) Capillary column 5% of phenyl-methylpolysiloxane. Oven temperature: 80°C (2min) increased to 230°C with a ramp of 15°C/min and then to 320°C with a ramp of 50°C/min, where it was held for 4 min. The injector and interface temperatures were maintained at 280°C.	n.a.	2.5	(K. Mohamed 2017)
Amphetamine, Methamphetamine, 3,4-methylenedioxyamphetamine, 3,4-methylenedioxymethamphetamine, 3,4-methylenedioxy-N-ethylamphetamine	Forehead skin surface sweat	Device cotton pad (0.5×0.5 cm)	HS-SPME (100 μm polydimethylsiloxane) and Derevatization with acetic anhydride and MSTFA	GC-MS (EI) Capillary column 5% phenylmethylsiloxane. Oven temperature: 60°C for 2 min, then raised 20°C/min to reach 250°C and finally held at 250°C for 8 min. The temperatures of the injection port, ion source and transfer line were set at 250°C, 230°C and 280°C, respectively.	0.27 ng/pad (Amphetamine); 0.21 ng/pad (Methamphetamine); 0.13 ng/pad (3,4-methylenedioxyamphetamine); 0.09 ng/pad (3,4-methylene-dioxymethamphetamine);	0.81 ng/pad (Amphetamine); 0.63 ng/pad (Methamphetamine); 0.39 ng/pad (3,4-methylenedioxyamphetamine); 0.27 ng/pad (3,4-methylenedioxymethamphetamine);	(Gentili et al. 2016)

Compounds	Sample	Volume/weight (mL or mg)	Extraction procedure	GC Conditions	LOD (ng/mL)	LOQ (ng/mL)	Ref.
					0.15 ng/pad (3,4-methylenedioxy-N-ethylamphetamine)	0.35 ng/pad (3,4-methylenedioxy-N-ethylamphetamine)	
Paramethoxy-amphetamine	Hair Blood Urine	10 for hair samples and 1 for blood and urine samples	LLE (ethylacetate) and Derivatization with TFAA	GC-MS Capillary column HP-5MS Oven temperature was set to 100°C for 1 min. The oven temperature was increased to 160°C at a rate of 15°C/min and maintained at 160°C for 3 min. Next, the temperature was increased to 280°C at a heating rate of 30°C/min and held for 10 min. The injector temperature was 250°C.	0.25 ng/mg for hair samples; 5 ng/mL for blood and urine samples	0.5 ng/mg for hair samples; 50 ng/mL for blood and urine samples	(Jang et al. 2016)

Legend: DLLME: Dispersive liquid liquid microextraction; EI: Electron ionization; GC-MS: Gas chromatography coupled mass spectrometry; GC-MS/MS: Gas chromatography coupled mass spectrometry in tandem; HFAA: heptafluorobutyric anhydride; HS-SPME: Headspace solid-phase microextraction; LLE: Liquid-liquid extraction; LOD: Limit of detection; LOQ: Limit of quantitation; MBTFA: N-methyl-bis-trifluoroacetamide; MEPS: Microextraction by packed sorbent; MS: Mass spectrometry; MSTFA: N-Methyl-N-trimethylsilyl-trifluoroacetamide; n.a.: not available; SPE: Solid-phase extraction; TFAA: Trifluoroacetic anhydride.

In the absence of the LOQ value, the lowest point of the calibration curve was considered.

Gentili et al. (Gentili et al. 2016), for sweat samples stored in device cotton pad, determined this class of compounds with LOD between 0.09 ng/pad and 0.27 ng/pad and LOQ values between 0.27 ng/pad and 0.81 ng/pad. The results obtained support the potential of sweat for non-invasive monitoring of exposure/intoxication by abused drugs in drivers, with advantages such as simple sample collection, easy pre-treatment and a viable economic alternative to conventional tests.

Concerning paramethoxyamphetamine, Jang et al. (Jang et al. 2016) developed a methodology for its determination in blood, urine and hair specimens, obtaining from 10 mg of the latter specimen a LOD of 0.25 ng/mg and a LOQ of 0.5 ng/mg. Hair is a valuable sample for retrospective analysis in drug abuse cases. In this paper, paramethoxyamphetamine was identified and quantified in *postmortem* specimens, which makes this method useful for the interpretation of results in forensic scenarios.

Additionally for *postmortem* situations, Ishikawa et al. (Ishikawa et al. 2018) developed a methodology for the determination of amphetamine derivatives in samples of 1 mL of vitreous humor. With treatment of these samples by LLE, they have reached LODs between 1 ng and 2.5 ng/mL and LOQs of 10 ng/mL. The authors concluded that the results demonstrate that methylenedioxyamphetamine derivatives can be detected at low concentrations and that they remain stable in this matrix at low temperatures for a period ranging from 8 hours to 5 weeks. Therefore, this method applied to samples of vitreous humor, can be used on routine forensic analysis for real case scenarios.

In the last year, Jarque et al. (Jarque et al. 2018) developed a method for the identification of amphetamine and methamphetamine in urine and hair samples but also in specimens of maternal and neonate origin (urine, hair and meconium), in order to understand the extent of prenatal exposure to these illicit compounds. For the hair samples (24-60 mg), the authors obtained a LOD of 1 ng/mg and a LOQ of 2 ng/mg. For 0.4 g of meconium they obtained 50 ng/g of LOD and 100 ng/g of LOQ. The data provided may be useful for clinical purposes to improve the diagnosis and monitoring of acute and chronic intoxications of mothers and newborns.

The volatility of these analytes requires derivatization. In addition, this procedure is necessary in order to improve the chromatographic peak and the mass spectral quality. Since they are polar or ionic, alkalinization or acidification is required to reduce the polarity of these drugs in order to improve the extraction phase/water partition coefficients of the analytes (Zeng et al. 2011). Alkyl chloroformates are the most commonly used reagents for the derivatization of amphetamine and derivatives in biological fluids (Mercieca et al. 2018; K. Mohamed 2017) because they produce water stable alkylcarbamates as they can react directly and rapidly in alkaline aqueous samples at room temperature and under stirring (Nakamoto et al. 2010; Mariotti et al. 2014). Anhydrides such as HFBA (Ishikawa et al. 2018; Jarque et al. 2018) and TFAA (Jang et al. 2016) are reagents usually used for acylation of the amino group of amphetamines (Dobos, Hidvégi, and Somogyi 2012; K. M. Mohamed and Bakdash 2017). In addition, with the derivatizing agent called MSTFA in mixture, more symmetric peaks are obtained for these analytes (Gentili et al. 2016; Chalavi et al. 2019). The use of hydroxylamine hydrochloride is less commun. This derivatization process is similar to the reaction of amines and oximes have also proven effective for drugs that contain ketones (Kerrigan 2015).

After derivatization, biological samples with amphetamines and derivatives can be analysed by gas chromatography. Mass spectrometers are the most commonly used detectors for the identification and quantification of these compounds, and all studies in table 4 were developed using GC-MS. This detector is preferably used for its advantage in separating the compounds under study by molecular weight, even if they present similar structures, and for its specificity, selectivity and sensitivity. Another detector used is the FID, which is more easily acquired and maintained by clinical and forensic laboratories.

An example of the use of this detector is the work developed by Xiong et al. (Xiong et al. 2010), who developed a methodology for the determination of amphetamine, methamphetamine, MDA and MDMA in 3 mL of urine samples. Through headspace-LPME pre-treatment of the specimens, the authors obtained LOD values between 8 and 82 ng/mL. Djozan et al.(Djozan et al. 2012), developed a method for the identification of methamphetamine and MDMA also in urine samples and detection by GC-FID. With molecularly extraction imprinted-solid phase extraction-DLLME, obtained LOD values between 2 and 18 ng/mL and a LOQs between 8 and 50 ng/mL.

Chalavi et al. (Chalavi et al. 2019) carried out a complete compilation of recent advances in microextraction procedures for determination of amphetamines and derivatives in biological samples and on the published analytical methods.

DETERMINATION OF NEW PSYCHOACTIVE SUBSTANCES

Psychoactive substances have been used for different purposes since the early times of History. However, their consumption now encompasses a wider range of substances than in the past, and as time passes, more and more substances are getting out to the drug market worldwide. Indeed, in the last decade a wide range of New Psychoactive Substances (NPS) have been appearing in the market, which only proves the constant evolution of the illicit drug production and consumerism (European Monitoring Centre for Drugs and Drug Addiction 2018).

Quoting both the Directive (EU) 2017/2103 and the European Monitoring Centre for Drugs and Drug Addiction (EMCDDA), NPS are new narcotic or psychotropic substances, in pure form or in preparation, not controlled by the Single Convention on Narcotic Drugs of 1961 nor the Convention on Psychotropic Substances of 1971 that can pose a public health risk to those posed by the substances listed in said conventions (European Monitoring Centre for Drugs and Drug Addiction 2018; Brendon Hughes, Ana Gallegos 2011; European Monitoring Centre for Drugs and Drug Addiction 2019).

These new emerging drugs may be divided into categories according to their structure or biological activity. Synthetic cannabinoids are the most common NPS, followed by cathinones, but opioids, benzodiazepines, arylcyclohexylamines and phenethylamines are also known NPS (European Monitoring Centre for Drugs and Drug Adiction 2019).

These substances are mainly fabricated as alternatives for controlled drugs that already exist (reason why they are also known as 'legal highs') (Zuba 2012; C. Margalho et al. 2016). Synthetic cannabinoids were synthetized as legal substitutes for cannabis and cathinones for MDMA and cocaine, mimicking or amplifying their effects (Baumann 2016). As NPS are added to the lists of controlled compounds, variants of them are being created that escape the legislation. However, NPS are not just new compounds; some substances were already known and used for some intents but are just now emerging in the illicit drug market for recreational use (Zuba 2012; United Nations Office on Drugs and Crime 2018; Dargan and Wood 2010)

These products are often sold as "harmless to health" substances, like bath salts, incense sticks or even fertilizers, usually containing labels such as "not suitable for human consumption" (Zuba 2012; Valente et al. 2014). Therefore, and because many times the labelling of these NPS does not correspond to their contents, their correct use is difficult which makes them potentially dangerous to public health (Ledberg 2015). These substances are usually synthesized using inexpensive low purity reagents and the final product is not subject to any quality control (Maciów-Głąb et al. 2014). Also, in contrast to new drugs, NPS are placed on the market without their pharmacological, pharmacokinetic and toxicological parameters being studied, or without corresponding analytical information (Zuba 2012; Maciów-Głąb et al. 2014). It is known, however, that they are not only addictive, but their consumption can cause permanent damage to consumers since they mainly act on the central nervous system (Maciów-Głąb et al. 2014; World Health Organization 2016), and there have been already several cases of fatal intoxications related to NPS consumption (Valente et al. 2014; Kuś et al. 2017).

In fact, the Directive (EU) 2017/2103 has recently revised the definition of "drug" and ten former NPS were added to the list of drugs considering their risk assessment. Any NPS can be now added to this list if it poses a severe public health risk (European Monitoring Centre for Drugs and Drug Addiction 2018).

Thus, and especially considering the latter, the monitoring of these compounds is essential but the penalties for their possession may vary depending on the NPS and country (Karila et al. 2018; Odoardi, Romolo, and Strano-Rossi 2016). Therefore, it is extremely important to consider the work of international institutions such as the EMCDDA (in Europe), responsible for monitoring and regulating this problem. As previously mentioned, the commercialization of NPS has been growing, which is due not only to technological development, but also to the low prices of reagents and the use of the internet for marketing and sale (Zuba 2012).

Therefore, forensic laboratories must be equipped with identification and quantification methods for analysing these compounds and their metabolites (Tiago Rosado, Gonçalves, et al. 2018), so that the correct legal sanctions can be applied. The immunoassays usually used for drug screening are not adequate for these substances, considering their structure diversity, their non-specificity and the insufficient cut-off levels (Lehmann et al. 2017). Also, and since new structures are entering the drug market at alarming levels, keeping an updated database of NPS standards turns out to be a continuous and complicated task, reason why there is a lack of analytical information regarding these substances.

It is therefore important to use sensitive hyphenated techniques such as GC-MS. Several analytical methodologies have been used for the identification and quantification of both synthetic cannabinoids and cathinones in biological samples, such as GC-MS, LC-MS, LC-TOF and LC-HRMS.

When it comes to the use of GC-MS, since many compounds can have the same retention time, it is important to associate gas chromatography with the ability to obtain mass spectra. The GC-MS combination is also vital to distinguish between similar structures and isomers (Zuba 2012). This analytical technique is fundamental for the identification of these substances considering the low concentrations at which they normally exist in samples, and most of the active components of NPS can be characterized by GC-MS due to their volatile properties (McMaster 2008; Lakshmi HimaBindu, Angala Parameswari, and Gopinath 2013).

Considering this technique, the use of EI mode is essential since it produces a high number of ions, thus contributing to high fragmentation and thereby demonstrating high selectivity. This mode allows obtaining information about the structure of the molecule based on their main fragmentation patterns in the MS spectrum. Both synthetic cannabinoids and cathinones have probable fragmentation pathways by EI (Namera et al. 2015), as demonstrated in Figure 1 A and B, respectively. The fragmentation of each compound is characteristic, so either by comparison with spectrum databases or by deducting the initial structure of the compounds themselves (when known prior to ionization), a correct identification of each compound is possible (Karila et al. 2018; Hübschmann 2009).

However, the molecular ion in the EI mass spectrum can be low in abundance or even undetectable. Split mode is also recommended when analysing NPS, since it is used when the analyte constitutes more than 0.1% of the mixture, with only a small percentage of the sample being analysed by GC-MS. The use of this mode provides cleaner results, which leads to a better understanding of the sample composition (Harris 2007).

Also, sensitivity is an important criterion when assessing the quality of detectors. The higher the sensitivity, the more likely it is that low detection limits can be obtained, and NPS are often found at very low concentrations in biological matrices. Thus, the use of a mass spectrometric detector is mandatory, considering its higher sensibility when compared to the so-called "universal detectors" (Hübschmann 2009).

It is also advised the use of the quadrupole analyser when dealing with these new substances (Hübschmann 2009; Batey 2014). The analysis can also be done either in full scan or in SIM modes. By using SIM, the sensitivity is higher, as only specific ion signatures are sought, so that it is possible to obtain lower detection limits. However, when dealing with samples in which the composition is unknown, full scan mode is recommended (Hübschmann 2009; Murray et al. 2013; IUPAC 2019).

In both GC-MS and LC-MS, sample preparation is necessary, especially when dealing with complex biological matrices such as blood. Pre-treatment of the sample is therefore necessary to remove unwanted matrix components, thereby reducing their effect and increasing sensitivity

(Mercieca et al. 2018). This procedure provides better results with reduced background noise and allows the preconcentration of the analytes, obtaining lower limits of detection. Therefore, extraction methods like SPE, LLE, PPT and DLLME can be used (Tiago Rosado, Gonçalves, et al. 2018).

The choice of the sample preparation method depends mainly on the biological specimen, which should also be chosen depending on the case (Gallardo and Queiroz 2008; Drummer 2004; Wiergowski et al. 2017). Blood and urine are the most common biological matrices analysed when dealing with NPS. Oral fluid (Tiago Rosado, Gonçalves, et al. 2018; Gallardo and Queiroz 2008; Gallardo, Barroso, and Queiroz 2009a), vitreous humor, pericardial fluid, hair, liver and gastric contents (Drummer 2004; Niu et al. 2018; Mário Barroso et al. 2010; Cláudia Margalho et al. 2016) may also be analysed. However, all biological matrices present advantages and disadvantages. Blood analysis allows establishing a correlation between the determined concentrations of NPS and their toxic effects on the body, but blood is also the most complex biological matrix, needing a stronger cleaning process prior to its analysis. Besides, some synthetic cannabinoids have a short half-life in whole blood samples, so its analysis should be used only in cases of acute intoxication. On the other hand, urine studies normally provide cleaner results, but the detection of NPS in this matrix is not always possible because some NPS are quickly biotrastransformed into metabolites that are not yet characterized.

On another note, the analysis of hair samples provides a wider detection window and reveals greater stability (Tiago Rosado, Gonçalves, et al. 2018; Salomone et al. 2017). Thus, the sample preparation process needs to be chosen accordingly. As previously mentioned, SPE is a very commonly used technique for sample preparation. Mixed-mode SPE is advised since it involves reverse phase and ion exchange: several different solvents can be used to remove interferences, but the analyte will always be retained by at least one mechanism (or both) (Sigma-Aldrich 2019). Thus, interferences may be retained by one mechanism, but only the analyte will be retained by the second mechanism. One of the main applications of this type of SPE is the isolation of drugs and metabolites from urine and blood samples (Thurman and Mills 1998).

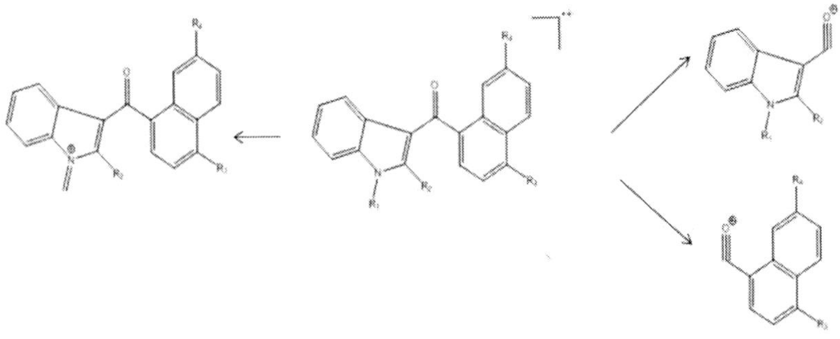

(A)

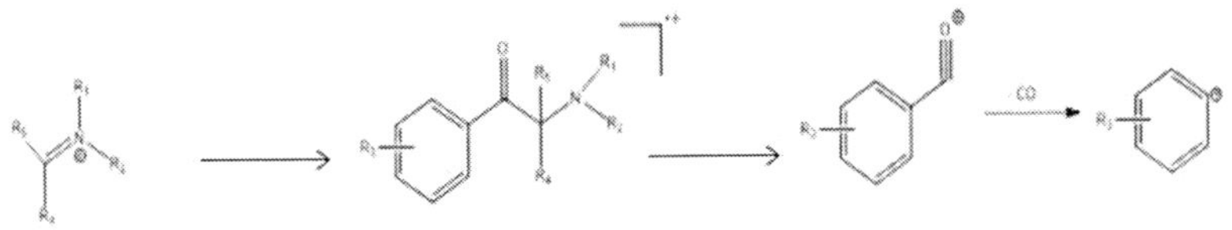

(B)

Figure 1. (A): Probable fragmentation of synthetic cannabinoids by EI; (B): Probable fragmentation of cathinones by EI.

It is also important to consider the need to use derivatization when analysing these new substances. This process can improve the resolution of compounds eluting at the same retention time and overlapping peaks, thereby increasing sensitivity, selectivity and resolution (Restek 2019). For instance, methylone, a very common cathinone, has functional groups in its structure, so it is normally analysed by GC-MS after proper derivatization. Especially when dealing with cathinones, derivatization modes using silylayion and acylation mechanisms are usually utilized, as they are effective reactions for primary and secondary amines.

Regarding the internal standard, it is advised the use of the equivalent analyte in deuterated form.

The use of GC-MS is progressively being used in forensic toxicology laboratories for the analysis of several types of NPS, and its use is providing an advance on the ongoing work of creating updated databases for analytical information regarding NPS.

THE USE OF GC IN THERAPEUTIC DRUG MONITORING

When referring to scientific fields such as toxicology and therapeutic drug monitoring (TDM), one must be aware that these are related, and both are based on the determination and quantification of many drugs and metabolites in biological specimens. These quantitative methods applied to samples such as blood, plasma and oral fluid demonstrate the great importance of TDM in clinical and toxicological laboratory analyses. The aim of this monitoring is to maximize some effects of particular drugs and minimize their toxic effects, so the therapeutic concentration is suitable and achieved according to the drugs specifications. Medicines that are conducive to monitoring usually have narrow therapeutic windows in which exist relationships between the observed effect and the bloodstream concentration (Snozek, Langman, and Cotten 2019). Many studies have been developed for several classes of drugs over the years, and the common methodologies developed for their quantification were based on immunoassays. In recent years, with the emergence of chromatography-based methods, TDM gained a new dimension and significance, since these methods enabled more sensitive, specific and rapid results. Although LC-MS/MS is currently the leading standard in a wide range of TDM studies, GC-MS is a highly selective technique and has sufficient sensitivity for the use in routine measurements (Madej and Kościelniak 2008). The latter technique is now widely used to determine with precision and accuracy several drugs (H. H. Maurer 2018).

In the last ten years, the number of studies carried out using GC has gradually increased and simultaneously new sample extraction techniques have been introduced to quantitative methods, which greatly contributed for the evolution of TDM. Table 5 reports several studies conducted in recent years, in which different drugs are determined with GC for TDM purposes.

Some examples concerning the main drugs that can be monitored by GC will be revised below.

In the particular case of antipsychotic drugs (APDs), some drugs present numerous adverse effects and, in most cases, monitoring of prescribed doses is advisable (Mandrioli et al. 2012). In the last few years, the monitoring of atypical APDs has been a subject of debate due to the improvement of extraction techniques applied to different samples and analytical methods. This improvement allowed the gathering of enough evidence to justify its relevance in patient compliance and treatment effectiveness (Grundmann, Kacirova, and Urinovska 2014). Regarding GC, not all available APDs are easily identified and quantified, olanzapine being one example. However, in a study by Ikeda et al. (Ikeda et al. 2012) olanzapine was determined in plasma samples by GC-MS and their method was successfully applied to three authentic samples. The authors revealed that the study was only a small step towards the therapeutic monitoring of olanzapine administration in patients and should be extended to several cases. In addition, Markowitz and Patrick (Markowitz and Patrick 1995) undertook studies to determine if clozapine-N-oxide, the main urinary metabolite of clozapine, could interfere with the GC-MS analysis of the parent drug. The authors observed a significant conversion on-column of clozapine-N-oxide to clozapine after injection onto a 5% (phenyl)-methylpolysiloxane capillary column operated at 250°C. In this sense, it was advised that preparation of biological specimens to determine clozapine by GC should avoid conditions which reportedly co-extract the N-oxide in order to avoid this artifact.

It is important to consider that most APDs are not volatile, so the preferred separation method for this class is LC (Zhang, Jr, and Bartlett 2008; Patteet et al. 2015). Nevertheless, for some APDs, general GC methodologies for TDM can be used as simple, less expensive and sensitive analytical procedures using either MS or NPD (Patteet et al. 2015; Zhang, Jr, and Bartlett 2008; Watelle et al. 1997; C. S. de la Torre, Martínez, and Almarza 2005). Bianchetti and Morselli (Bianchetti and Morselli 1978) developed a sensitive GC-NPD method to determine haloperidol in human plasma and proved its suitability for routine analysis of haloperidol plasma levels in patients undergoing this treatment. The authors used 2.0 mL of plasma sample and LLE as purification technique, achieving a LOQ of 1.0 ng/mL. McKay et al. (McKay et al. 1982) used a GC-MS procedure for the determination of chlorpromazine in plasma and compared it with a HPLC coupled to an electrochemical detector system. LLE was applied as well, achieving a LOQ of 0.25 ng/mL. Later, Hattori et al. (Hattori, Suzuki, and

Brandenberger 1986) reported a GC-MS procedure for the determination of haloperidol in both plasma and urine samples. Additionally, the authors presented the mass spectra of this APD with EI, PCI and NCI. Later on, de la Torre et al. (C. S. de la Torre, Martínez, and Almarza 2005) developed a reliable GC-NPD method to determine chlorpromazine, levomepromazine, olanzapine, clozapine and haloperidol in whole blood specimens. SPE was used as purification and pre-concentration of the APDs, obtaining LOQs from 0.1 to 0.5 ng/mL.

One way to improve the separation of several APDs, increasing its volatility and the method sensitivity is derivatization. A GC-MS method for the determination of clozapine and norclozapine was described using TFAA as derivatization reagent and other method derivatized amisulpride with HFBA (Patteet et al. 2015). On the other hand, the trends in mass spectrometry with the introduction of MS/MS also enabled the improvement on sensitivity and specificity of the developed methods. According to a review by Patteet et al. (Patteet et al. 2015), between 2010 and 2014 only four GC–MS/MS methods for determination of antipsychotics had been published. A MEPS cleanup procedure coupled to GC–MS/MS to determine chlorpromazine, haloperidol, cyamemazine, quetiapine, clozapine, olanzapine and levomepromazine in 0.25 mL of plasma specimens (Da Fonseca et al. 2013) was presented. The authors used MSTFA with 5% TMS as derivatization reagent. The obtained LOQs ranged from 0.2 to 1 ng/mL. More recently, Rosado et al. (Tiago Rosado, Oppolzer, et al. 2018) demonstrated the importance of quantifying several APDs in plasma and oral fluid samples using GC-MS/MS. In fact, this was the first study reporting the determination of this class of drugs in oral fluid using MS/MS. The authors used a volume of 0.2 mL of sample and obtained LODs that ranged from 1 to 10 ng/mL. They further stated that the developed method could establish APDs levels correlations between the two specimens (plasma and oral fluid), also contributing for TDM with the advantage of oral fluid presenting less invasive collection. Following the previous study, Caramelo et al. (Caramelo et al. 2019) also developed a GC-MS/MS method for the quantification of several APDs in oral fluid samples considering as the most prominent factors, the small sample volume used (0.05 mL) and the advantages of the extraction procedure applied. The authors used a sampling approach named dried saliva spots (DSS), which proved to be a great alternative for routine analysis compared to other classic procedures.

GC is also commonly used for the determination of antidepressant drugs. In addition, GC becomes a simple, high-resolution, sensitive, reproducible and inexpensive method to quantify this class of drugs in several matrices (Grundmann, Kacirova, and Urinovska 2014).

Antidepressant drugs (ADs) are the first line of treatment for depression and other psychological disorders, and their use has been increasing worldwide. The increasing prescription of ADs received special attention, and the search for more sensitive and selective methods to quantify these drugs levels became an important contribution for TDM (Mandrioli et al. 2012).

It is accepted that not all ADs can be quantified by GC either due to chromatographic (trazodone) or derivatization problems (O-desmethylvenlafaxine) (Wille et al. 2007). In addition, desipramine, and nortriptyline are particularly susceptible to the aging of liners and columns, leading to a progressive loss of sensitivity (Drummer et al. 1994). Suzuki, et al. (Suzuki et al. 1986) reported EI, CI and NCI mass spectra of some ADs using direct inlet system, but they did not take into account that several drugs (e.g., N-oxides) are decomposed either on sample preparation or GC (Pfleger, Maurer, and Weber 1992; H. H. Maurer 1992).

However, Wille et al.(Wille et al. 2007) developed a method that can determine most ADs in the therapeutic range using EI. The authors state that PCI and NCI lead to higher selectivity, with NCI being of outmost interest for small sample volumes and high sensitivity requirements. Maurer and Pfleger (H. Maurer and Pfleger 1984) described in 1984 a GC-MS screening in urine using packed columns, and later confirmed that modern capillary columns lead to the same or better results (H. H. Maurer 1992), and GC retention indexes corresponded to those measured on capillary columns (De Zeeuw 1992).

Regarding tricyclic antidepressants (TCAs), GC is highly selective and presents enough sensitivity for routine measurements, pharmacokinetic studies and forensic analysis depending on the detector (Norman and

Maguire 1985). The FID is considered the detector with lowest sensitivity and relatively large volumes of sample are needed to be extracted to ensure adequate sensitivity and detection limits (Norman and Maguire 1985). The first practical FID analysis of TCAs was reported for amitriptyline and nortriptyline (R. Braithwaite and Widdop 1971; R. A. Braithwaite and Whatley 1970). A coated GC packing phase 3% OV-17 gave satisfactory resolution for tertiary and secondary amine drugs, and the use of derivatization reagents, such as acetic or trifluoroacetic anhydride, significantly improved the tailing of secondary amine peaks (Norman and Maguire 1985). In addition, a GC-FID method was proposed by Berzas Nevado et al. (Berzas Nevado et al. 2000) for the simultaneous determination of clomipramine, fluoxetine and fluvoxamine in pharmaceutical formulations without pre-derivatization.

The coupling of GC to electron-capture detection (ECD) is well suited to all applications of antidepressant measurements, allowing lower LODs and greater sensitivity. However, it is somewhat technically more difficult and has the limitation of the necessity to convert drugs into a species which readily capture electrons, usually halogenated compounds (Norman and Maguire 1985). This is readily achieved for secondary amine drugs by conversion to a trifluoroacetyl or heptafluorobutyryl derivative, but regarding the tertiary amine antidepressants the analysis procedure can be more complex (Norman and Maguire 1985). The transformation of the tertiary amines into the respective carbamate by reaction with several chloroformate derivatives has successfully been applied on several TCAs determinations, producing a compound with good electron capturing characteristics.

The NPD also reveals outstanding sensitivity for the detection of traces of drugs and negligible interference from non-nitrogenous compounds, both endogenous and exogenous. It shows enough sensitivity for the determination of this class of ADs and is a powerful tool concerning underivatized drug analysis (Martínez, De La Torre, and Almarza 2003).

Two SPE procedures were evaluated for the simultaneous determination of seven ADs in whole blood, including amitriptyline, nortriptyline, trimipramine and clomipramine using GC-NPD and LOQs of 25, 51, 37 and 223 ng/mL were reported respectively (Madej and Kościelniak 2008; Martínez, De La Torre, and Almarza 2003). A simultaneous quantitative determination of amitriptyline, nortriptyline, imipramine, desipramine, clomipramine and desmethylclomipramine in human plasma was performed by GC–NPD by de la Torre et al. (R. de la Torre et al. 1998), obtaining LODs that ranged from 1.2 to 5.8 ng/ml. Also, Ulrich and Martens (Ulrich and Martens 1997) performed an assay for ten ADs in human plasma or serum after SPME with GC-NPD, reaching a LOD of 10 ng/mL.

Nowadays, mass spectrometric detection provides a very sensitive approach for drug measurements and has been widely applied to ADs. Nevertheless, TCAs undergo extensive fragmentation at high ionisation energies, resulting in a similar mass spectrum with a major peak at m/z of 58 due to loss of part of the side-chain, and, as previously reported by Wille et al. (Wille et al. 2007), CI may be chosen since greater structural integrity is retained in the spectrum (Norman and Maguire 1985). Shen et al. (Shen et al. 2002) studied the determination of TCAs present at therapeutic levels in hair samples.

The drugs and metabolites were identified using GC-MS using EI and CI, but their quantification was performed using GC-NPD. The authors found concentrations of 57.7 ng/mg for amitriptyline and 183.3 ng/mg for doxepin, 68.2 ng/mg for chlorpromazine, and 36.8 ng/mg for trifluoroperazine (Madej and Kościelniak 2008). Other GC-MS methods were developed for the simultaneous determination and identification of TCAs, such as imipramine, desipramine, amitriptyline, amoxapine, doxepin, trimipramine and metabolites, in human plasma (X. Lee et al. 2008; Pommier, Sioufi, and Godbillon 1997) and hair (Sporkert and Pragst 2000). However, Pommier et al. (Pommier, Sioufi, and Godbillon 1997) had to convert desipramine into its pentafluoropropionyl derivative before detection.

These TCAs were detected using either EI or PCI, with a LOD of 0.2–0.5 ng/mL for fluid specimens, and 0.05–1.0 ng/mg for hair (Uddin, Samanidou, and Papadoyannis 2011). Way et al. reported an isotope dilution GC-MS measurement of TCAs in plasma and showed the utility of the 4-carbethoxyhexa-fluorobutyryl derivatives of secondary amines, achieving a LOD of 8 ng/mL (Uddin, Samanidou, and Papadoyannis 2011; Way et al. 1998).

Pujadas et al. developed a simple and reliable GC-MS method for the determination of TCAs in oral fluid and derivatization by MSTFA. The authors used EI ionization mode and obtained a LOQ of 0.9 ng/mL (Pujadas et al. 2007). In most of the described methods, a fused silica capillary column coated with cross-linked dimethyl-, diphenyl- or phenyl methyl siloxane were used to carry out TCAs separation, and helium was the most commonly used carrier gas (Uddin, Samanidou, and Papadoyannis 2011).

Another important class of ADs is the selective serotonin reuptake inhibitors (SSRIs), which differ widely in their chemical structure (Eap and Baumann 1996). Methods for quantification of SSRIs were reviewed in 1996 by Eap and Baumann (Eap and Baumann 1996). Simultaneous GC-MS quantification was recently described by the same working group (Eap et al. 1998).

Previously, the available methods for fluoxetine and norfluoxetine were based on achiral GC-ECD. The enantiomers of fluoxetine and norfluoxetine can be determined following derivatization with (S)-trifluoroacetylprolyl chloride and separation achieved on an achiral GC column (DB-5 crosslinked fused silica capillary) (Torok-Both et al. 1992). The first among several available methods to determine fluvoxamine was developed using derivatization and analysis by GC-ECD (Hurst et al. 1981). Also, GC-ECD methods have been described for sertraline and desmethylsertraline (Tremaine and Joerg 1989). Paroxetine can be assayed in human plasma using GC-NPD (Petersen et al. 1978).

As with TCAs, GC-MS is the most used technique for SSRIs in present days. This is also justified by the fact that full-scan EI mode is considered a reference technique for drug screening procedures and provides comparable GC-MS spectra on all commercially available instruments due to uniform mass fragmentation patterns (Unceta, Goicolea, and Barrio 2011). Eap et al. (Eap et al. 1998) presented a GC-MS method which allowed the simultaneous determination of citalopram, paroxetine, sertraline, and their active N-demethylated metabolites (desmethyl-citalopram, didesmethyl-citalopram, and desmethylsertraline) after derivatization with MBTFA. The authors reported LOQs of 2 ng/mL for citalopram and paroxetine, 1 ng/mL for sertraline, and 0.5 ng/mL for desmethylcitalopram. A SPME–GC–MS method was proposed by Salgado-Petinal et al. (Salgado-Petinal et al. 2005) for the determination of the frequently prescribed venlafaxine, fluvoxamine, mirtazapine, fluoxetine, citalopram, and sertraline in urine samples with LODs lower than 0.4 ng/mL.

In this method fluvoxamine, fluoxetine, and sertraline were determined as their acetyl derivatives, while venlafaxine, mirtazapine, and citalopram were determined unchanged. More recently, Rosado et al. (Tiago Rosado, Gonçalves, Martinho, et al. 2017) optimized and validated a method for the determination of several SSRIs amongst other ADs, including metabolites, in urine and plasma samples using GC–MS. The authors used MSTFA as derivatization reagent and reached LOQs that varied from 1 to 15 ng/mL for both specimens. These authors made an important study on the cross-contribution that occurred for the different studied ADs. This cross contribution was observed by the presence of chromatographic peaks at the retention times of non-injected ADs when extracting its ions on SIM mode. The subsequent decision to split the ADs working standard solutions in two different mixtures had already been applied by Wille et al. (Wille et al. 2007).

This would eliminate the effect of contribution or at least minimize it to below 5% (Tiago Rosado, Gonçalves, Martinho, et al. 2017). Rosado et al. (Tiago Rosado, Gonçalves, Martinho, et al. 2017) also observed the extensive fragmentation of the ADs chosen for the multimethod, which resulted in a similar mass spectrum, and consequently did not allow the application of MS/MS detection. Previously, a study by Déglon et al. (Déglon et al. 2010) used DBS as sample extraction technique to quantify fuoxetine and its major metabolite (norfluoxetine), reboxetine and paroxetine with GC-NCI-MS/MS. The authors used a small sample volume (10 µL) and were able to reach LOQs of 1 ng/mL for fluoxetine and its metabolite, and 20 ng/mL for reboxetine and paroxetine. Papoutsis et al. (Papoutsis et al. 2012) presented a GC-MS method for the quantification of eleven ADs, including SSRIs, and four of their metabolites in whole blood. The authors used SPE to preconcentrate the analytes and achieved LOQs ranging from 1 to 5 ng/mL, concluding its great applicability for TDM (Papoutsis et al. 2012).

The most recent study in this field was reported by Feng et al. (Feng et al. 2019), that demonstrated the determination of four ADs in plasma and urine by GC-MS combined with sensitive and selective derivatization. The ADs were derivatized with sodium nitrite to appropriate N-nitrosamines under acidic condition, allowing easier detection of the derivatives by GC-MS, obtaining LOQs of 0.14 to 4.62 ng/mL.

Other group of drugs that could be monitored by GC are those used to treat and/or prevent seizures (antiepileptic drugs, AEDs, also known as anticonvulsant drugs), and have been among the most common drugs for which TDM is performed (McMillin and Krasowski 2016). Apart from multiple reasons that make TDM useful, the main reason that stands out in these specific drugs is the interindividual variability that antiepileptic pharmacokinetics demonstrates.

Several chromatographic methods exist that support TDM of phenytoin, including GC with FID (Ritz, Gerald Warren, and Teitelbaum 1975). Other procedures for the determination of underivatized phenytoin have been reported, but unfortunately they suffer from the same drawbacks as barbiturates (Burke and Thenot 1985). Tailing peaks can frequently occur unless great care is taken to obtain a deactivated column. Nevertheless, it is possible to obtain symmetrical peaks with the new stationary phases or with cross-linked capillary columns (Burke and Thenot 1985).

However, the most widely applied procedures to determine hydantoins involve derivatization. Procedures that incorporate methyl and trimethylsilyl derivatization techniques are known to improve its sensitivity (Abraham and Joslin 1976; Chang and Glazko 1970; Kupferberg 1970). Additional methods include GC with NPD and MS detection (Nelson et al. 1998; Vandemark and Adams 1976).

Carbamazepine and its metabolite are known to be unstable under most GC conditions (Juenke and McMillin 2009). It was early realized that this AED is unstable in the injection port of the chromatograph and degrades to iminostilbene and to 9-methylacridine. The same happens with the epoxide metabolite rearranging to 9-acridine carboxy aldehyde (Burke and Thenot 1985). However there are many methods that use preanalytical derivatization and have proven useful in the analysis of carbamazepine (Juenke and McMillin 2009). These include dimethylformamide demethylacetal (Millner and Taber 1979; Perchalski and Wilder 1974) pentafluorobenzamide (Schwertner, Hamilton, and Wallace 1978), N-cyano (Gérardin, Abadie, and Laffont 1975), and trimethyl sylil (Lensmeyer 1977) derivates. Underivatized evaluations demonstrated a variable decomposition to iminostilbene (Cocks, Dyer, and Edgar 1981) or use MS detection (Hallbach, Vogel, and Guder 1997). The epoxide metabolite of carbamazepine is also easily degraded by GC, and for this reason the most popular approach for the simultaneous determination of both compounds is HPLC (Juenke and McMillin 2009).

Regarding valproic acid, the most commonly employed chromatographic technique is GC, directly using FID (Tosoni, Signorini, and Albertini 1983; Wohler and Poklis 1997; Yu and Shih 1996), an ester derivative (Calendrillo and Reynoso 1980; Morita et al. 1981; Nishioka et al. 1985), or by MS (Darius and Meyer 1994). This AED can be determined as well as the methyl (Willox and Foote 1978; Gyllenhaal and Albinsson 1978; Tupper, Solow, and Kenfield 1978), propyl (Morita et al. 1981), butyl (Hulshoff and Roseboom 1979) or phenacyl (Gupta, Eng, and Gupta 1979; Rege et al. 1984; Chan 1980) derivatives. Chromatography of valproic acid does not create any specific problem. Most methods are based on the separation with columns for free fatty acids (FFAP columns or 10% SP-

1000) (Burke and Thenot 1985). Wohler and Poklis (Wohler and Poklis 1997) determined valproic acid in serum without prior derivatization on a Nukol wide-bore capillary column and reported LOD and LOQ of 5 and 10 µg/mL, respectively. However, Darius and Meyer (Darius and Meyer 1994) described a sensitive GC-MS method for the determination of valproic acid and 7 of its metabolites. The method was based on the SIM analysis of the tert-butyldimethylsilyl derivatives using MTBSTFA as derivatization reagent. The obtained LOD for valproic acid and metabolites was in the low ng/mL range, except for the 4-hydroxy metabolite with a LOD of 100 ng/mL. Nau et al. (Nau et al. 1981) proposed a GC-MS procedure for the determination of valproic acid and its metabolites as TMS derivatives. The authors reported a sensitivity limit of 3-6 ng/mL for most metabolites with a sample size of 200 µL.

Also, chromatographically, phenobarbital is most often analysed by GC, generally using FID or NPD (Vandemark and Adams 1976). This AED is a weak acid (pKa, 7.3), and both nitrogens N-1 and N-3 may be replaced by an alkyl group. It is therefore amenable to GC determination either unchanged or as an alkyl derivative (Burke and Thenot 1985). Analysis of the underivatized molecule and on-column alkylation of phenobarbital are by far the most often utilized procedures in routine practice (Burke and Thenot 1985). The power of GC-MS becomes more evident in metabolic studies requiring the determination of metabolites along with the parent drug. The disposition of most barbiturates, and of phenobarbital in particular, has been assayed with GC-MS techniques (Kapetanović and Kupferberg 1980; Goldberg et al. 1979; Van et al. 1982).

Typically monitored by GC methods are also methsuximide and its active metabolite. A simple GC-NPD assay for the analysis of methsuximide, its metabolite and ethoitin has been described (Juenke and McMillin 2009). Regarding GC-MS, methsuximide and its N-demethylated metabolite, can both be determined underivatized under EI conditions using the fragment obtained by the loss of the imide group (STRONG et al. 1974), or following butylation (Burke and Thenot 1985).

Nowadays, alternative specimens such as oral fluid are getting special attention for TDM of AEDs, although its applicability is yet limited. The reason of this great interest is due to an ease of sample collection and non-invasive sampling. Nevertheless, there are some studies in which AEDs have been determined in this matrix (Krasowski 2010). In a study by Hosli et al. (Hosli et al. 2013), phenotoin was not only quantified in saliva, but also in other samples such as blood and brain microdialysate. The authors used GC-MS and achieved LODs of 15 ng/mL and LOQs of 50 ng/mL. Mecarelli et al. (Mecarelli et al. 2007) carried out a study of correlation between saliva and serum concentrations for levetiracetam. Samples were extracted by SPE, analysed by GC-MS and a LOD below 10 ng/mL was obtained. The authors concluded that saliva could be an alternative to serum for the determination of this drug. On the other hand, two studies presented new extraction techniques for valproic acid with the same concept, DBS and dried matrix spots (DMS) for blood and plasma, respectively (Rhoden et al. 2014; Ikeda et al. 2014). Both studies used GC-MS, but Rhoden et al. (Rhoden et al. 2014) achieved a lower LOQ than that of the work by Ikeda et al., (Ikeda et al. 2014). The different specimens used might justify the results obtained by the authors.

Over the years, TDM has expanded to all classes of drugs. As example, Athanasiadou et al. (Athanasiadou et al. 2014) quantified busulfan, an antineoplastic agent. The monitoring of this agent is frequently required in conditioning regimens prior to bone marrow transplantation for the treatment of leukemia and other types of cancer. As it happens with the previously described classes, this drug has wide inter- and intra-individual variability, which makes TDM of plasma levels mandatory in clinical practice.

The developed method used GC-MS and aimed to demonstrate good accuracy and specificity as well as a simple and fast sample preparation process for bulsufan monitoring in clinical practice (Athanasiadou et al. 2014). Furthermore, Hasegawa et al. (Hasegawa et al. 2011) developed and validated a GC-MS method for the quantification of dextromethorphan in plasma samples. This drug is an antitussive available without a prescription, which makes it interesting for TDM since it reveals hallucinogenic effects, euphoria and psychosis when used at high doses.

Table 5. Review of GC-methods for the quantification of different drugs with application in TDM

Compounds	Sample	Volume (mL)	Extraction procedure	GC conditions	LOD (ng/mL)	LOQ (ng/mL)	Ref.
Chlorpromazine, levomepromazine, cyamemazine, clozapine, haloperidol quetiapine	Oral fluid	0.05	DSS and derivatization with MSTFA with 5% TMCS.	GC-MS/MS Capillary column (30 m × 0.25 mm i.d., 0.25 µm film thickness) with a 5% phenylmethylsiloxane (HP-5MS). The initial oven temperature was held ate 120°C for 2 min, then raised to 300°C at a 20°C/min rate (held for 14 min). The mass spectrometer operated with a filament current of 35 µA and electron energy of 70 eV in the positive EI mode. The injection temperature was set to 250°C and the temperature of ion source was 280°C.	n.s.	5-10	(Caramelo et al. 2019)
Chlorpromazine, levomepromazine, cyamemazine, clozapine, haloperidol, olanzapine, quetiapine	Plasma/ Oral fluid	0.5 of plasma and 0.2 of oral fluid	SPE (Strata™-X-C). Derivatization with MSTFA with 5% TMCS.	GC-MS/MS Capillary column (30 m × 0.25 mm i.d., 0.25 µm film thickness) with a 5% phenylmethylsiloxane (HP-5MS). The initial oven temperature was held at 150°C during 2 min, then raised to 300 °C at 40°C/min (held for 8.5 min). The mass spectrometer was operated with a filament current of 35µA and electron energy of 70 eV in EI mode. The temperatures of injection port and ion source were, respectively, 250°C and 280°C.	Plasma: 1- 10 Oral fluid: 1- 5	Plasma: 2 - 40 Oral fluid: 2 - 10	(Tiago Rosado, Oppolzer, et al. 2018)

Compounds	Sample	Volume (mL)	Extraction procedure	GC conditions	LOD (ng/mL)	LOQ (ng/mL)	Ref.
Olanzapine	Plasma	0.5	Silylation of Pyrex® glass tubes. LLE (dichloromethane/ n-hexane (1:1, v/v)). Addition 50 μL of ethyl acetate containing 1% triethylamine was added.	GC-MS. Capillary column (30 m × 0.25 mm i.d., 0.25 μm film thickness) equipped with a fused-silica (DB-5MS). The oven temperature was programmed to 80°C and held for 2 min, and then increased from 80 to 300°C at a rate of 15°C/min. The mass spectrometer was operated with EI mode.	n.a.	0.5	(Ikeda et al. 2012)
Fluoxetine, nortriptyline, maprotiline, paroxetine	Plasma/ Urine	2.0	LLE (dichloromethane). Derivatization with hydrochloric acid and saturated sodium nitrite solution.	GC-MS. Capillary column (30 m × 0.25 mm i.d., 0.25 μm film thickness) with 5% phenylmethylsiloxane (HP-5MS). The oven temperature was programmed 130°C, raised to 240°C at a rate of 10°C/min (held for 2 min), then increased to 250°C at a rate of 8°C/min and finally raised to 260°C and held for 10 min in a rate of 12°C/min. The mass spectrometer was operated at EI (70 eV) mode. The temperatures of injection port and ion source were 250°C and 230°C, respectively.	0,04 - 1,38	0,14 - 4,62	(Feng et al. 2019)

Table 5. (Continued)

Compounds	Sample	Volume (mL)	Extraction procedure	GC conditions	LOD (ng/mL)	LOQ (ng/mL)	Ref.
Levetiracetam, lamotrigine	Whole blood	0.2	SPE (HF Bond Elut C18 columns). Derivatization with acetonitrile and MTBSTFA with 1% TBDMSCl.	GC-MS Capillary column DB-5MS (30 m × 0.25 mm i.d., 0.25 μm film thickness). The oven temperature started in 70°C (held for 1 min), increased to 120°C at a rate of 10°C/min, and then raised to 300°C at a rate of 40°C/min, held for 6.5 min. The mass spectrometer was performed in EI mode. The temperatures of injection port, ion source and interface were 240, 200 and 300°C, respectively.	150	500	(Nikolaou, Papoutsis, Dona, et al. 2015)
Valproic acid	Whole blood	0.05	DBS	GC-MS Capillary column CP-WAX (30 m × 0.25 mm i.d., 0.25 μm). The column temperature was held at 80°C for 2 min and then increased at a rate of 40°C/min until reaching 250°C, then it was maintained for 2 min. The temperatures of transfer line, injection and ion source were 280°C, 250°C and 300°C, respectively.	n.a.	500	(Rhoden et al. 2014)

Compounds	Sample	Volume (mL)	Extraction procedure	GC conditions	LOD (ng/mL)	LOQ (ng/mL)	Ref.
Valproic acid, gabapentin	Plasma	0.02	DMS and Derivatization with MSTFA.	GC-MS Capillary column DB-5 MS (30m × 0.25 mm i.d., 0.25 µm film thickness). The oven temperature was programmed to 50°C (held for 1 min) and then increased to 300°C at a rate of 20°C/min. The mass spectrometer was operated in EI mode.	n.a.	500 - 10000	(Ikeda et al. 2014)
Valproic acid, salicylic acid, paracetamol, phenobarbital, primidone, phenytoin	Plasma	0.2	450 µL of butyl acetate was added to the plasma samples and addition of acetic acid.	GC-MS Capillary column fused silica HP-1MS (12 m × 0.25 mm i.d., 0.25 µm film thickness). The initial oven temperature was set at 85°C for 3 min and raised to 310°C, held for 1.57 min, at a rate of 35°C/min. The mass spectrometer was operated in EI mode with a ionization energy of 70 eV and the temperature of ion source was 240°C.	n.a	2×10^6 - 1×10^8	(Meyer et al. 2011)
Lacosamide	Plasma	0.2	2.0 mL of phosphate buffer 0.1 M (pH 6.00) and SPE (HF Bond Elut LRC-C18). Derivatization by silylation (30 µL of acetonitrile and 30 µL of MTBSTFA with 1% TBDMSCl)	GC-MS Capillary column DB-5MS (30 m × 0.25 mm i.d., 0.25 µm film thickness). The conditions of oven temperature were: 70°C initially held for 1 min, increased at a rate of 10°C/min to 120°C and finally reached to 300°C at a rate of 40°C/min (held for 6.5 min). The temperatures of ion source, interface and injection were 200, 300 and 240°C respectively. The mass spectrometer was performed in EI mode.	60	200	(Nikolaou, Papoutsis, Spiliopoulou, et al. 2015)

Table 5. (Continued)

Compounds	Sample	Volume (mL)	Extraction procedure	GC conditions	LOD (ng/mL)	LOQ (ng/mL)	Ref.
Busulfan	Plasma	0.5	LLE with ethyl acetate. Reconstitution with 100 μL I_2 solution in acetonitrile (0.25%, w/v).	GC-MS Capillary column (12 m × 0.200 mm i.d., 0.33 μm film thickness) with 5% phenyl-methylpolysiloxane split fused silica (HP-Ultra 2). The gradient of temperatures were as follows: initial temperature of 80°C, increased to 170°C at a rate of 15°C/min and then ramped at 40°C/min to 310°C (held to 2 min). The temperatures of ion source, transfer line and front inlet of the detector were 230, 300 and 250°C respectively. The mass spectrometer was operated at EI mode and the ionization energy was 70 eV.	10.6	25	(Athanasiadou et al. 2014)
Phenytoin	Brain microdialysate / Oral fluid/ Blood	0.5	SPE (C8-SCX). Reconstitution and derivatization with 50 μL of TMSH.	GC-MS Capillary column DB-5MS (30 m × 0.25 mm i.d., 0.25 μm film). The oven temperature was set to 120°C maintained for 1 min, raised to 300°C at a rate of 10°C/min and held for 6 min. The temperature of the ion source was 230°C and the transfer line was 280°C.	15	50	(Hosli et al. 2013)

Compounds	Sample	Volume (mL)	Extraction procedure	GC conditions	LOD (ng/mL)	LOQ (ng/mL)	Ref.
Oxcarbazepine, carbamazepine, phenytoin, alprazolam	Plasma/ Urine	1000 of plasma and 10000 of urine	Centrifugation and MEPS (4 mg C18) material.	GC-MS Capillary column Rtx-1 MS (30 m × 0.25 mm i.d., 0.25 µm film). The column oven temperature was hold for 1 min at 100°C, then increased to 200°C at a rate of 10°C/min, raised from 200°C to 260°C at a rate of 15°C/min and finally reached to 300°C by 30°C/min. The temperatures for the injector and the ion source were 270°C. The mass spectrometer was performed at EI mode and the ionization energy was 70 eV.	Plasma: 0.0018 – 0.0036 Urine: 0.0020 – 0.0029	Plasma: 0.0056 – 0.0108 Urine: 0.0060 – 0.0088	(Rani and Malik 2012)
Amitriptyline, citalopram, clomipramine, fluoxetine, fluvoxamine, maprotiline and metabolite, mirtazapine and metabolite, nortriptyline, paroxetine, sertraline and metabolite, venlafaxine and metabolite	Whole blood	1.0	Addition of 4 mL of phosphate buffer, followed by centrifugation and then SPE (Bond Elut LRC Certify). Acylation with HFBA in 100 µL ethyl acetate, at 50 °C for 30 min.	GC-MS Capillary column HP- 5MS (30 m × 0.25 mm i.d., 0.25 µm film thickness). The parameters for oven temperature were as follows: 100°C was held for 1 min, then increased to 300°C at a rate of 40°C/min and was held for 4 min. The injection temperature was 280 and the transfer line temperature was 300°C. The mass spectrometer was operated with 70 eV and in the EI mode.	0.30 – 1.50	1.00 – 5.00	(Papoutsis et al. 2012)

Table 5. (Continued)

Compounds	Sample	Volume (mL)	Extraction procedure	GC conditions	LOD (ng/mL)	LOQ (ng/mL)	Ref.
Fluoxetine and metabolite Reboxetine Paroxetine	Whole blood	0.01	DBS and derivatization with 0.02% triethylamine in butyl chloride and PFPA.	GC-NICI-MS/MS Capillary column DB-5MS (15 m × 0.25 mm i.d., film thickness 0.25 μm). The oven temperature was set at 105°C for 1 min, raised to 300°C at 70°C/min (held for 1.22 min). The temperatures of injection, transfer line and ion source were 300, 275 and 150°C, respectively. The mass spectrometer was operated in the negative-ion chemical ionisation (NICI) mode.	0.02 for all compounds	1.0–20	(Déglon et al. 2010)
Levetiracetam	Oral fluid/serum	1,00	SPE (Bond Elut LRC Certify) and derivatization with MSTFA with 2% TMCS.	GC-EI-MS Capillary column (25 m × 0.20 mm i.d., film thickness 0.33 μm) with 95% dimethyl–5% diphenyl polysiloxane (HP-5MS). The oven temperature initially was set from 110°C to 280°C at a rate of 20°C/min (held 1 min). Temperatures of the injection and transfer line were both 280°C. The ionization of the mass spectrometer was set at 70 eV and in EI mode.	< 10	n.s.	(Mecarelli et al. 2007)

Compounds	Sample	Volume (mL)	Extraction procedure	GC conditions	LOD (ng/mL)	LOQ (ng/mL)	Ref.
Dextromethorphan	Plasma	0.1	300 μL of distilled water and 50 μL of glycine-sodium hydroxide buffer (1 mol/L) were added to the samples and after the centrifugation the MonoTip C18 tip extraction was applied.	GC-EI-MS Capillary column Equity-5 fused silica (30 m × 0.32 mm i.d., film thickness 0.25 μm). The GC conditions for the oven temperature were as follows: initially set to 120°C for 1 min, then at 20°C/min to 270°C and finally raised to 300°C at 30°C/min. The interface temperature was 300°C, the ion source was 250°C and the injection was operated at 200°C. The mass spectrometer was performed by EI mode with an ionization energy of 70 EV and emission current of 60 μA.	1.25	2.5	(Hasegawa et al. 2011)

Legend: DBS: Dried blood spots; DMS: Dried matrix spots DSS: Dried saliva spots; EI: Electron ionization; GC-MS: Gas chromatography coupled mass spectrometry; GC-MS/MS: Gas chromatography coupled mass spectrometry in tandem; HFBA: heptafluorobutyric anhydride; HS-SPME: Headspace solid-phase microextraction; LLE: Liquid-liquid extraction; LOD: Limit of detection; LOQ: Limit of quantitation; MBTFA: N-methyl-bis-trifluoroacetamide; MTBSTFA: N-tert-Butyldimethylsilyl-N-methyltrifluoroacetamide; MEPS: Microextraction by packed sorbent; MSTFA: N-Methyl-N-trimethylsilyl-trifluoroacetamide; n.a.: not available; PFPA: pentafluoropropionic anhydride; SPE: Solid-phase extraction; TBDMSCl: tert-butyldimethylsilylchloride; TFAA: trifluoroacetic anhydride

In the absence of the LOQ value, the lowest point of the calibration curve was considered.

It is important to highlight that although LC remains the most commonly chromatographic method used for TDM, its costs and operation makes many researchers explore other alternatives. One of those is GC, particularly coupled to MS, which shows multiple advantages in the quantification of several drugs. The combination of GC with simplified extraction methods is a great solution for the disadvantages of LC, allowing greater applicability for TDM and analysis in forensic and clinical toxicology.

METABOLOMICS

Metabolomics involves the determination of the complete set of metabolites present in cells, body fluids and tissues, usually known as the metabolome (Koek et al. 2011). It becomes a great challenge to develop generic methodologies for the analysis of a complete metabolome, or at least the possible majority of the present metabolites, according to the metabolome complexity (Koek et al. 2011).

In a toxicological analysis, the first step is the identification of unknown drugs in body fluids. However, there might be several problems resulting from a "general unknown" analysis (H. H. Maurer 1992). There are thousands of drugs that could have been consumed, and each drug may form several metabolites, turning identification more complicated (H. H. Maurer 1992). Some drugs suffer a complete metabolism so that can only be identified in plasma or urine by their metabolites. On the other hand, all metabolites must be differentiated from other potential compounds, and often are present at low concentrations (H. H. Maurer 1992). Apart from toxicological analysis, most metabolomics fields of applications are disease diagnosis (Claudino et al. 2012; Wikoff et al. 2007), biomarker screening (Xue et al. 2008; Bogdanov et al. 2008), and characterization of biological pathways (Nicholson et al. 2002). There are two traditional ways of metabolomic studies approaching: (i) targeted metabolomics, a quantitative analysis of one or more pre-selected metabolites, and (ii) untargeted metabolomics, that assumes the high throughput analysis of the metabolic state contained in the biological system (biological fluid or cell culture) by quantification of the concentration profile of all possible metabolites (Sawada et al. 2008; Dunn et al. 2013).

Nowadays, the widely used analytical techniques applied to a metabolome are nuclear magnetic resonance spectroscopy (NMR) and hyphenated alternatives such as GC or LC coupled to MS (Koek et al. 2011).

GC-MS is well suited for a comprehensive analysis, combining the high separation efficiency with the versatile, selective and sensitive MS detection (Koek et al. 2011). This technique is far from ideal for metabolomics, as it is limited to drugs that are either volatile or can be made volatile through derivatization; and all nonvolatile drugs must be carefully extracted from the sample before analysis, requiring demanding sample treatment. But on the other hand, it is consensual that no single analytical technique is able to detect an entire set of metabolites in a biological specimen (Garcia and Barbas 2011).

In fact, GC-MS has been used for metabolite profiling decades before LC-MS (Kind et al. 2009; Horning and Horning 1971; Thompson and Markey 1975), primarily due to the difficult ionization process in LC-MS demanding target analytes to be ionized, separated from the liquid solvents, and funneled into the MS for analysis (Kind et al. 2009). In addition, LC-MS commonly uses electrospray ionization which is considered softer than the GC-MS EI ionization, and generates for the most part molecular ion adducts with little or no in-source fragmentation (Sawada et al. 2008). For this reason, most of GC-based applications for metabolomics purposes combine GC with MS using EI ionization, which has been standardized at 70 eV in the late 1960s (Koek et al. 2011; Kind et al. 2009). The latter is justified by the fact that full scan response in EI mode is approximately proportional to the amount of compound injected and that the identification of peaks through an extensive available mass spectra database is straightforward, due to the reproducible fragmentation patterns obtained in this mode. Furthermore, the fragmentation pattern can be of extreme importance for the identification and/or classification of unknown metabolites (Koek et al. 2011).

Bando et al. (Bando et al. 2011) lead a study with a GC-MS-based metabolomics approach for toxicological evaluation, and tried to elucidate the mechanism of toxicity of hydrazine using metabolic profiles obtained from the analysis. Hydrazine is a metabolite of some drugs such as the antihypertensive drug hydralazine and the antituberculosis drug isoniazid. It has been reported that hydrazine induces hepatotoxicity, carcinogenicity, mutagenicity, teratology and neurotoxicity. The metabolome analysis was performed via non-target approach and the authors observed a metabolic profiling with dose-dependent toxicity induced by hydrazine. In other application field, Wu et al. (H. Wu et al. 2009) presented a method to investigate the urinary metabolic difference between hepatocellular carcinoma male patients and normal male subjects. The urinary endogenous metabolome was assayed using derivatization with BSTFA and 1% TMCS followed by GC-MS and a total of 103 metabolites were detected, of which 66 were known compounds. The authors concluded that this non-invasive technique of identifying biomarkers from urine may have clinical utility. Later, the same authors studied the difference of metabolomic profile between normal and malignant gastric tissue in order to further explore tumor biomarkers (H. Wu et al. 2010). The authors used again derivatization with BSTFA plus 1% TMCS together with GC-MS to obtain the metabolomic information of the malignant and non-malignant tissues of gastric mucosae in gastric cancer patients. The obtained results showed that 18 detected metabolites were different between the malignant tissues and the adjacent non-malignant tissues of gastric mucosa. Additionally, 5 detected metabolites were also different between the non-invasive and invasive tumors. The authors concluded that the selected tissue metabolites could be applied for clinical diagnosis or staging for gastric cancer (H. Wu et al. 2010). Chen et al. (J.-L. Chen et al. 2010), aimed at elucidating the underlying mechanisms of metastasis and at identifying the metabolomic markers of gastric cancer metastasis. The analysis of the gastric carcinoma tissue samples was performed by GC-MS, using as derivatization reagent N-methyl-N-t-butyldimethyl-silyltrifluoro-acetamide (MBDSTFA) and it was possible to observe that 29 metabolites were differentially expressed in animal models of human gastric cancer. Of those metabolites, 20 were up-regulated and 9 were down-regulated in metastasis group compared to the non-metastasis group. The authors concluded that the metabolic profiling of tumor tissues can provide new biomarkers for the treatment of gastric cancer metastasis. Xue et al. (Xue et al. 2008) investigated the serum metabolic difference between hepatocellular carcinoma male patients and normal male subjects. There were detected 13 metabolites that discriminate between the hepatocellular carcinoma patients and healthy subjects, and this detection was performed with the help of derivatization with MSTFA and 1% TMCS followed by GC-MS.

Furthermore, GC-MS analysis of the urinary metabolome is also very important to study mutations of inborn errors of metabolism (Kuhara 2005). Tanaka et al. (Tanaka et al. 1966) described a new genetic defect of leucine metabolism, in which a marked accumulation of isovaleric acid, a catabolite of leucine, occurred. The obtained GC-MS data clearly identified the abnormal metabolite as isovaleric acid. The authors observed that two studied siblings with this hereditary disease were found to have an intolerance to protein and developed recurrent episodes of metabolic acidosis, with stupor or coma. Hommes et al. (Hommes et al. 1968) described propionicacidemia a new inborn error of metabolism. The authors observed that the hepatic fat of a five-year-old deceased child contained an increased amount of C_{15} and C_{17} saturated fatty acids, probably resulting from a disturbance in the utilization of these fatty acids at the level of propionic acid. The analysis was performed by GC to the methyl esters of the fatty acids, and differences were observed. This metabolome analysis, either by urease pre-treatment of urine or eluates from dried urine on filter-paper, stable-isotope dilution, and GC-MS with full-scan and extracted ion chromatograms has allowed the simultaneous molecular diagnosis of numerous inborn errors of metabolism (Kuhara 2005).

Predominantly, derivatization is used for GC-based metabolomics, and Kanani and Klapa (H. H. Kanani and Klapa 2007) pointed out potential sources of derivatization biases. These authors reported that when a derivatization reaction is required there are three quantities of interest that need to be seriously considered for an accurate data acquisition, quantification and analysis process: (i) the concentration of a metabolite in the original specimen, (ii) the concentration of the metabolite derivative(s) in the derivatized specimen, and (iii) the measured peak area(s) of the derivative(s) (H. H. Kanani and Klapa 2007). Considering the last two as the same can lead to erroneous data acquisition and analysis protocols. In addition, silylation with derivatizing reagents such as BSTFA or MSTFA are conducted commonly for metabolic profiling, especially for biofluid and tissue sample metabolomic researches (Xu, Zou, and Ong 2009). Nevertheless, in order to reduce or eliminate conversion reactions during silylation, methoxamine hydrochloride can first be used for a prior oximation reaction (Lindon, Nicholson, and Holmes 2011; Viant 2008; Issaq et al. 2009; Shulaev 2006; Fiehn 2008; Pasikanti, Ho, and Chan 2008; H. Kanani, Chrysanthopoulos, and Klapa 2008; Xu, Zou, and Ong 2009). This additional step usually leads to a more complex chromatogram associated to a large background noise (Xu, Zou, and Ong 2009).

Although conventional GC has been demonstrated to provide an adequate resolution for a variety of analytical tasks, it might not resolve the numerous components present in the studied specimens (Almstetter, Oefner, and Dettmer 2012). Nowadays, great applications have been performed in the metobolomics field with 2D-GC. This instrumentation shows superior chromatographic resolving power and becomes particularly suitable for the separation of low-molecular-weight analytes in complex matrices (Liu and Phillips 1991; Mondello et al. 2008; Almstetter, Oefner, and Dettmer 2012). Briefly, a (thermal or pressure-based) modulator locates between columns for a periodic focus of the effluent from one and transfer it to the other in small concentrated segments (Bertsch 2000; Almstetter, Oefner, and Dettmer 2012). This creates narrow second dimension (2D) peaks, increasing peak heights and subsequently enhances the sensitivity (Górecki, Harynuk, and Panić 2004; Almstetter, Oefner, and Dettmer 2012; Bertsch 2000). Lately, the coupling of 2D-GC with TOF became one of the preferred detection techniques (Almstetter, Oefner, and Dettmer 2012). Welthagen et al. (Welthagen et al. 2005) used 2D-GC-TOF for high resolution metabolomics, in which discovered biomarkers on spleen tissue extracts of obese NZO compared to lean C57BL/6 mice. The authors report an improved peak capacity with 2D-GC allowing the detection of peaks that could not previously be separated in one-dimensional GC. Caldeira et al. (Caldeira et al. 2012) also used 2D-GC-TOF in the analysis of the exhaled breath from 32 children with allergic asthma, 10 of which also had allergic rhinitis, and 27 control children. The authors found a pattern of 6 compounds belonging to the alkanes group that characterized the asthmatic population, and highlight that a compound set was successfully applied with possible clinical applications.

Metabolomic sciences have advanced over the last decades and partially due to the combination of different and increasingly sensitive techniques. The GC-MS-based metabolomics will continue to be a useful tool for different research fields. Increasing databases, along with computational methods to predict metabolite identification, will be essential for this progress. Furthermore, with the acquired knowledge from metabolomics and with the support of increasingly sensitive and ecofriendly extraction /purification techniques applied to specimens, it will be possible to improve the diagnostic of certain diseases.

FUTURE PERSPECTIVES AND CONCLUSION

Nowadays, despite the enormous advances in liquid chromatographic techniques associated to mass spectrometry, GC systems coupled to several mass spectrometric detectors continues relevant to the analytical field. The robustness of this technique, sensitivity and accessibility for most laboratories continue to be the main advantages of using this instrumentation, despite the disadvantages associated to the often time-consuming derivatization processes. However, different microwave assisted derivatization methods have been developed that make this process no longer a disadvantage. Today, most of GC methods for the determination of drugs of abuse in most forensic toxicology laboratories are routine procedures to identify different drugs of abuse (scan mode), for the identification of new substances as well as for quantification purposes. Undoubtedly, the new challenges that are going to arise are associated with the application of this instrumentation in the area of metabolomics, which is why the relevant publications associated with the use of GC coupled to hybrid systems will certainly be in this field.

ACKNOWLEDGMENTS

This work is supported by FEDER funds through the POCI - COMPETE 2020 - Operational Programme Competitiveness and Internationalisation in Axis I - Strengthening research, technological development and innovation (Project POCI-01-0145-FEDER-007491) and National Funds by FCT - Fundação para a Ciência e a Tecnologia (Project UID/Multi/00709/2019). T. Rosado acknowledges the Centro de Competências em Cloud Computing in the form of a fellowship (C4_WP2.6_M1 – Bioinformatics; Operação UBIMEDICAL – CENTRO-01-0145-FEDER-000019 – C4 – Centro de Competências em Cloud Computing, supported by Fundo Europeu de Desenvolvimento Regional (FEDER) through the Programa Operacional Regional Centro (Centro 2020). S. Soares and J. Gonçalves acknowledge the FCT in the form of fellowships (SFRH/BD/148753/2019) and (SFRH/BD/149360/2019), respectively.

REFERENCES

Abraham, Chempithera Varughese, and Harry Douglas Joslin. 1976. "Simultaneous Gas Chromatographic Analysis for the Four Commonly Used Antiepileptic Drugs in Serum." *Journal of Chromatography A* 128 (2): 281–87.

Aizpurua-Olaizola, Oier, Jone Omar, Patricia Navarro, Maitane Olivares, Nestor Etxebarria, and Aresatz Usobiaga. 2014. "Identification and Quantification of Cannabinoids in Cannabis Sativa L. Plants by High Performance Liquid Chromatography-Mass Spectrometry." *Analytical and Bioanalytical Chemistry* 406 (29): 7549–60. https://doi.org/10.1007/s00216-014-8177-x.

Aleksa, Katarina, Paula Walasek, Netta Fulga, Bhushan Kapur, Joey Gareri, and Gideon Koren. 2012. "Simultaneous Detection of Seventeen Drugs of Abuse and Metabolites in Hair Using Solid Phase Micro Extraction

(SPME) with GC/MS." *Forensic Science International* 218 (1–3): 31–36. https://doi.org/10.1016/j.forsciint.2011.10.002.

Almstetter, Martin F., Peter J. Oefner, and Katja Dettmer. 2012. "Comprehensive Two-Dimensional Gas Chromatography in Metabolomics." *Analytical and Bioanalytical Chemistry* 402 (6): 1993–2013.

Alvear, Eduardo, Dietrich Von Baer, Claudia Mardones, and Antonieta Hitschfeld. 2014. "Determination of Cocaine and Its Major Metabolite Benzoylecgonine in Several Matrices Obtained from Deceased Individuals with Presumed Drug Consumption Prior to Death." *Journal of Forensic and Legal Medicine* 23 (March): 37–43. https://doi.org/10.1016/j.jflm.2014.01.003.

Amaral, C., E. Gallardo, R. Rodrigues, R. Pinto Leite, D. Quelhas, C. Tomaz, and M.L. Cardoso. 2010. "Quantitative Analysis of Five Sterols in Amniotic Fluid by GC-MS: Application to the Diagnosis of Cholesterol Biosynthesis Defects." *Journal of Chromatography B: Analytical Technologies in the Biomedical and Life Sciences* 878 (23). https://doi.org/10.1016/j.jchromb.2010.06.010.

Andrews, Rebecca, and Sue Paterson. 2012. "A Validated Method for the Analysis of Cannabinoids in Post-Mortem Blood Using Liquid-Liquid Extraction and Two-Dimensional Gas Chromatography-Mass Spectrometry." *Forensic Science International* 222 (1–3): 111–17. https://doi.org/10.1016/j.forsciint.2012.05.007.

Angeli, Ilaria, Sara Casati, Alessandro Ravelli, Mauro Minoli, and Marica Orioli. 2018. "A Novel Single-Step GC–MS/MS Method for Cannabinoids and 11-OH-THC Metabolite Analysis in Hair." *Journal of Pharmaceutical and Biomedical Analysis* 155 (June): 1–6. https://doi.org/10.1016/j.jpba.2018.03.031.

Athanasiadou, Ioanna, Yiannis S. Angelis, Emmanouil Lyris, Helen Archontaki, Costas Georgakopoulos, and Georgia Valsami. 2014. "Gas Chromatographic–mass Spectrometric Quantitation of Busulfan in Human Plasma for Therapeutic Drug Monitoring: A New on-Line Derivatization Procedure for the Conversion of Busulfan to 1,4-

Diiodobutane." *Journal of Pharmaceutical and Biomedical Analysis* 90 (March): 207–14. https://doi.org/10.1016/J.JPBA.2013.12.004.

Balikova, M., V. Maresova, and V. Habrdova. 2001. "Evaluation of Urinary Dihydrocodeine Excretion in Human by Gas Chromatography-Mass Spectrometry." *Journal of Chromatography B: Biomedical Sciences and Applications* 752 (1): 179–86. https://doi.org/10.1016/S0378-4347(00)00509-0.

Bando, Kiyoko, Takeshi Kunimatsu, Jun Sakai, Juki Kimura, Hitoshi Funabashi, Takaki Seki, Takeshi Bamba, and Eiichiro Fukusaki. 2011. "GC-MS-based Metabolomics Reveals Mechanism of Action for Hydrazine Induced Hepatotoxicity in Rats." *Journal of Applied Toxicology* 31 (6): 524–35.

Barnes, Allan J., Karl B. Scheidweiler, and Marilyn A. Huestis. 2014. "Quantification of 11-nor-9-Carboxy-Δ9-Tetrahydrocannabinol in Human Oral Fluid by Gas Chromatography-Tandem Mass Spectrometry." *Therapeutic Drug Monitoring* 36 (2): 225–33. https://doi.org/10.1097/01.ftd.0000443071.30662.01.

Barroso, M., M. Dias, D. N. Vieira, M. López-Rivadulla, and J. A. Queiroz. 2010. "Simultaneous Quantitation of Morphine, 6-Acetylmorphine, Codeine, 6-Acetylcodeine and Tramadol in Hair Using Mixed-Mode Solid-Phase Extraction and Gas Chromatography-Mass Spectrometry." *Analytical and Bioanalytical Chemistry* 396 (8): 3059–69. https://doi.org/10.1007/s00216-010-3499-9.

Barroso, M., E. Gallardo, C. Margalho, E. Marques, D.N. Vieira, and M. López-Rivadulla. 2005. "Determination of Strychnine in Human Blood Using Solid-Phase Extraction and GC-EI-MS." *Journal of Analytical Toxicology* 29 (5).

Barroso, M., E. Gallardo, and J. A. Queiroz. 2009. "Bioanalytical Methods for the Determination of Cocaine and Metabolites in Human Biological Samples." *Bioanalysis*. Future Science Ltd. https://doi.org/10.4155/bio.09.72.

Barroso, M., E. Gallardo, D. N. Vieira, J. A. Queiroz, and M. López-Rivadulla. 2011. "Bioanalytical Procedures and Recent Developments in the Determination of Opiates/Opioids in Human Biological Samples."

Analytical and Bioanalytical Chemistry 400 (6): 1665–90. https://doi.org/10.1007/s00216-011-4888-4.

Barroso, Mário, and Eugenia Gallardo. 2015. "Assessing Cocaine Abuse Using LC-MS/MS Measurements in Biological Specimens." *Bioanalysis* 7 (12): 1497–1525. https://doi.org/10.4155/bio.15.72.

Barroso, Mário, Eugenia Gallardo, Duarte Nuno Vieira, Manuel López-Rivadulla, and João António Queiroz. 2010. "Hair: A Complementary Source of Bioanalytical Information in Forensic Toxicology." *Bioanalysis* 3 (1): 67–79. https://doi.org/10.4155/bio.10.171.

Batey, Jonathan H. 2014. "The Physics and Technology of Quadrupole Mass Spectrometers." *Vacuum* 101: 410–15. https://doi.org/10.1016/j.vacuum.2013.05.005.

Baumann, Michael H. 2016. "The Changing Face of Recreational Drug Use." *Cerebrum : The Dana Forum on Brain Science* 2016 (January): 1–15. http://www.ncbi.nlm.nih.gov/pubmed/27408674%0Ahttp://www.pubmedcentral.nih.gov/articlerender.fcgi?artid=PMC4938259.

Béres, Tibor, Lucie Černochová, Sanja Ćavar Zeljković, Sandra Benická, Tomáš Gucký, Michal Berčák, and Petr Tarkowski. 2019. "Intralaboratory Comparison of Analytical Methods for Quantification of Major Phytocannabinoids." *Analytical and Bioanalytical Chemistry* 411 (14): 3069–79. https://doi.org/10.1007/s00216-019-01760-y.

Bermejo, A. M., I. Ramos, P. Fernandez, M. Lopez-Rivadulla, A. Cruz, M. Chiarotti, N. Fucci, and R. Marsilli. 1992. "Morphine Determination by Gas Chromatography/Mass Spectroscopy in Human Vitreous Humor and Comparison with Radioimmunoassay." *Journal of Analytical Toxicology* 16 (6): 372–74.

Bertsch, Wolfgang. 2000. "Two-Dimensional Gas Chromatography. Concepts, Instrumentation, and Applications–Part 2: Comprehensive Two-Dimensional Gas Chromatography." *Journal of High Resolution Chromatography* 23 (3): 167–81.

Berzas Nevado, J. J., M. J. Villasenor Llerena, A. M. Contento Salcedo, and E. Aguas Nuevo. 2000. "Determination of Fluoxetine, Fluvoxamine, and Clomipramine in Pharmaceutical Formulations by Capillary Gas

Chromatography." *Journal of Chromatographic Science* 38 (5): 200–206.

Bianchetti, G., and P. L. Morselli. 1978. "Rapid and Sensitive Method for Determination of Haloperidol in Human Samples Using Nitrogen—phosphorus Selective Detection." *Journal of Chromatography A* 153 (1): 203–9.

Bogdanov, Mikhail, Wayne R. Matson, Lei Wang, Theodore Matson, Rachel Saunders-Pullman, Susan S. Bressman, and M. Flint Beal. 2008. "Metabolomic Profiling to Develop Blood Biomarkers for Parkinson's Disease." *Brain* 131 (2): 389–96.

Bowie, Lemuel J., and Peter B. Kirkpatrick. 1989. "Simultaneous Determination of Monoacetylmorphine, Morphine, Codeine, and Other Opiates by GC/MS." *Journal of Analytical Toxicology* 13 (6): 326–29.

Brabanter, Nik De, Wim Van Gansbeke, Fiona Hooghe, and Peter Van Eenoo. 2013. "Fast Quantification of 11-nor-Δ9-Tetrahydro-cannabinol-9-Carboxylic Acid (THCA) Using Microwave-Accelerated Derivatisation and Gas Chromatography–triple Quadrupole Mass Spectrometry." *Forensic Science International* 224 (1–3): 90–95. https://doi.org/10.1016/j.forsciint.2012.11.004.

Braithwaite, R. A., and B. Widdop. 1971. "A Specific Gas-Chromatographic Method for the Measurement of 'Steady-State' Plasma Levels of Amitriptyline and Nortriptyline in Patients." *Clinica Chimica Acta* 35 (2): 461–72.

Braithwaite, R. A., and J. A.. Whatley. 1970. "Specific Gas Chroma-tographic Determination of Amitriptyline in Human Urine Following Therapeutic Doses." *Journal of Chromatography A* 49: 303–7.

Brendon Hughes, Ana Gallegos, Roumen Sedefov. 2011. "Responding to New Psychoactive Substances." *Drugs in Focus, EMCDDA.*

Brunet, Bertrand R., Allan J. Barnes, Karl B. Scheidweiler, Patrick Mura, and Marilyn A. Huestis. 2008. "Development and Validation of a Solid-Phase Extraction Gas Chromatography-Mass Spectrometry Method for the Simultaneous Quantification of Methadone, Heroin, Cocaine and Metabolites in Sweat." *Analytical and Bioanalytical Chemistry* 392 (1–2): 115–27. https://doi.org/10.1007/s00216-008-2228-0.

Burke, J. T., and J. P. Thenot. 1985. "Determination of Antiepileptic Drugs." *Journal of Chromatography B: Biomedical Sciences and Applications* 340: 199–241.

Caldeira, M., R. Perestrelo, António S. Barros, M. J. Bilelo, A. Morete, J. S. Camara, and Sílvia M. Rocha. 2012. "Allergic Asthma Exhaled Breath Metabolome: A Challenge for Comprehensive Two-Dimensional Gas Chromatography." *Journal of Chromatography A* 1254: 87–97.

Caldwell, R., and H. Challenger. 1989. "A Capillary Column Gas-Chromatographic Method for the Identification of Drugs of Abuse in Urine Samples." *Annals of Clinical Biochemistry* 26 (5): 430–43.

Calendrillo, Bruce A., and Gustavo Reynoso. 1980. "A Micromethod for the On-Column Methylation of Valproic Acid by Gas-Liquid Chromatography." *Journal of Analytical Toxicology* 4 (6): 272–74.

Cámpora, P., A. M. Bermejo, M. J. Tabernero, and P. Fernández. 2003. "Quantitation of Cocaine and Its Major Metabolites in Human Saliva Using Gas Chromatography-Positive Chemical Ionization-Mass Spectrometry (GC-PCI-MS)." In *Journal of Analytical Toxicology*, 27:270–74. Preston Publications. https://doi.org/10.1093/jat/27.5.270.

Caramelo, Débora, Tiago Rosado, Victor Oliveira, Jesus M. Rodilla, Pedro M. M. Rocha, Mário Barroso, and Eugenia Gallardo. 2019. "Determination of Antipsychotic Drugs in Oral Fluid Using Dried Saliva Spots by Gas Chromatography-Tandem Mass Spectrometry." *Analytical and Bioanalytical Chemistry* 411 (23): 6141–53. https://doi.org/10.1007/s00216-019-02005-8.

Chalavi, Soheila, Sakine Asadi, Saeed Nojavan, and Ali Reza Fakhari. 2019. "Recent Advances in Microextraction Procedures for Determination of Amphetamines in Biological Samples." *Bioanalysis* 11 (5): 437–60. https://doi.org/10.4155/bio-2018-0207.

Chan, S. C. 1980. "Monitoring Serum Valproic Acid by Gas Chromatography with Electron-Capture Detection." *Clinical Chemistry* 26 (11): 1528–30.

Chang, Tsun, and A. J. Glazko. 1970. "Quantitative Assay of 5, 5-Diphenylhydantoin (Dilantin∗) and 5-(p-Hydroxyphenyl)-5-

Phenylhyda-ntoin by Gas-Liquid Chromatography." *The Journal of Laboratory and Clinical Medicine* 75 (1): 145–55.

Chen, Bai Hsiun, E. Howard Taylor, and Alex A. Pappas. 1990. "Comparison of Derivatives for Determination of Codeine and Morphine by Gas Chromatography/Mass Spectrometry." *Journal of Analytical Toxicology* 14 (1): 12–17.

Chen, Jin-Lian, Hui-Qing Tang, Jun-Duo Hu, Jing Fan, Jing Hong, and Jian-Zhong Gu. 2010. "Metabolomics of Gastric Cancer Metastasis Detected by Gas Chromatography and Mass Spectrometry." *World Journal of Gastroenterology: WJG* 16 (46): 5874.

Claudino, Wederson M., Priscila H. Goncalves, Angelo di Leo, Philip A. Philip, and Fazlul H. Sarkar. 2012. "Metabolomics in Cancer: A Bench-to-Bedside Intersection." *Critical Reviews in Oncology/Hematology* 84 (1): 1–7.

Cocks, D. A., T. F. Dyer, and K. Edgar. 1981. "Simple and Rapid Gas—liquid Chromatographic Method for Estimating Carbamazepine in Serum." *Journal of Chromatography B: Biomedical Sciences and Applications* 222 (3): 496–500.

Cognard, Emmanuelle, Stéphane Bouchonnet, and Christian Staub. 2006. "Validation of a Gas Chromatography-Ion Trap Tandem Mass Spectrometry for Simultaneous Analyse of Cocaine and Its Metabolites in Saliva." *Journal of Pharmaceutical and Biomedical Analysis* 41 (3): 925–34. https://doi.org/10.1016/j.jpba.2006.01.041.

Dargan, Paul, and David Wood. 2010. *Technical Report on Mephedrone, 2010. European Monitoring Centre for Drugs and Drug Addiction. Risk Assessment Report of a New Psychoactive Subsance: 4-Methylmethcathinone (Mephedrone)*.

Darius, Jörg, and Frank Peter Meyer. 1994. "Sensitive Capillary Gas Chromatographic—mass Spectrometric Method for the Therapeutic Drug Monitoring of Valproic Acid and Seven of Its Metabolites in Human Serum Application of the Assay for a Group of Pediatric Epileptics." *Journal of Chromatography B: Biomedical Sciences and Applications* 656 (2): 343–51.

Déglon, Julien, Estelle Lauer, Aurélien Thomas, Patrice Mangin, and Christian Staub. 2010. "Use of the Dried Blood Spot Sampling Process Coupled with Fast Gas Chromatography and Negative-Ion Chemical Ionization Tandem Mass Spectrometry: Application to Fluoxetine, Norfluoxetine, Reboxetine, and Paroxetine Analysis." *Analytical and Bioanalytical Chemistry* 396 (7): 2523–32. https://doi.org/10.1007/s00216-009-3412-6.

Djozan, Djavanshir, Mir Ali Farajzadeh, Saeed Mohammad Sorouraddin, and Tahmineh Baheri. 2012. "Molecularly Imprinted-Solid Phase Extraction Combined with Simultaneous Derivatization and Dispersive Liquid-Liquid Microextraction for Selective Extraction and Preconcentration of Methamphetamine and Ecstasy from Urine Samples Followed by Gas Chromatography." *Journal of Chromatography. A* 1248: 24–31. https://doi.org/10.1016/j.chroma.2012.05.085.

Dobos, Adrienn, Elod Hidvégi, and Gábor Pál Somogyi. 2012. "Comparison of Five Derivatizing Agents for the Determination of Amphetamine-Type Stimulants in Human Urine by Extractive Acylation and Gas Chromatography-Mass Spectrometry." *Journal of Analytical Toxicology* 36 (5): 340–44. https://doi.org/10.1093/jat/bks026.

Drost, R. H., R. D. Van Ooijen, T. Ionescu, and R. A. A. Maes. 1984. "Determination of Morphine in Serum and Cerebrospinal Fluid by Gas Chromatography and Selected Ion Monitoring after Reversed-Phase Column Extraction." *Journal of Chromatography B: Biomedical Sciences and Applications* 310: 193–98.

Drummer, Olaf H. 2005. "Review: Pharmacokinetics of Illicit Drugs in Oral Fluid." *Forensic Science International* 150 (2–3): 133–42. https://doi.org/10.1016/j.forsciint.2004.11.022.

Drummer, Olaf H. 2004. "Postmortem Toxicology of Drugs of Abuse." *Forensic Science International* 142 (2): 101–13. https://doi.org/https://doi.org/10.1016/j.forsciint.2004.02.013.

Drummer, Olaf H., Soumela Horomidis, Sophie Kourtis, Marie L. Syrjanen, and Patricia Tippett. 1994. "Capillary Gas Chromatographic Drug Screen for Use in Forensic Toxicology." *Journal of Analytical Toxicology* 18 (3): 134–38.

Dunn, Warwick B., Alexander Erban, Ralf J. M. Weber, Darren J. Creek, Marie Brown, Rainer Breitling, Thomas Hankemeier, Royston Goodacre, Steffen Neumann, and Joachim Kopka. 2013. "Mass Appeal: Metabolite Identification in Mass Spectrometry-Focused Untargeted Metabolomics." *Metabolomics* 9 (1): 44–66.

Eap, C. B., and P. Baumann. 1996. "Analytical Methods for the Quantitative Determination of Selective Serotonin Reuptake Inhibitors for Therapeutic Drug Monitoring Purposes in Patients." *Journal of Chromatography B: Biomedical Applications* 686 (1): 51–63. https://doi.org/10.1016/S0378-4347(96)00338-6.

Eap, C. B., G. Bouchoux, M. Amey, N. Cochard, L. Savary, and P. Baumann. 1998. "Simultaneous Determination of Human Plasma Levels of Citalopram, Paroxetine, Sertraline, and Their Metabolites by Gas Chromatography—Mass Spectrometry." *Journal of Chromatographic Science* 36 (7): 365–71.

Ellefsen, Kayla N., Marta Concheiro, Sandrine Pirard, David A. Gorelick, and Marilyn A. Huestis. 2016. "Oral Fluid Cocaine and Benzoylecgonine Concentrations Following Controlled Intravenous Cocaine Administration." *Forensic Science International* 260 (March): 95–101. https://doi.org/10.1016/j.forsciint.2016.01.013.

Ellefsen, Kayla N., Jose Luiz Da Costa, Marta Concheiro, Sebastien Anizan, Allan J. Barnes, Sandrine Pirard, David A. Gorelick, and Marilyn A. Huestis. 2015. "Cocaine and Metabolite Concentrations in DBS and Venous Blood after Controlled Intravenous Cocaine Administration." *Bioanalysis* 7 (16): 2041–56. https://doi.org/10.4155/bio.15.127.

ElSohly, Hala N., Don F. Stanford, A. B. Jones, M. A. ElSohly, H. Snyder, and C. Pedersen. 1988. "Gas Chromatographic/Mass Spectrometric Analysis of Morphine and Codeine in Human Urine of Poppy Seed Eaters." *Journal of Forensic Science* 33 (2): 347–56.

Emídio, Elissandro Soares, Vanessa de Menezes Prata, Fernando José Malagueño de Santana, and Haroldo Silveira Dórea. 2010. "Hollow Fiber-Based Liquid Phase Microextraction with Factorial Design Optimization and Gas Chromatography-Tandem Mass Spectrometry for Determination of Cannabinoids in Human Hair." *Journal of*

Chromatography. B, Analytical Technologies in the Biomedical and Life Sciences 878 (24): 2175–83. https://doi.org/10.1016/j.jchromb. 2010.06.005.

Emídio, Elissandro Soares, Vanessa de Menezes Prata, and Haroldo Silveira Dórea. 2010. "Validation of an Analytical Method for Analysis of Cannabinoids in Hair by Headspace Solid-Phase Microextraction and Gas Chromatography-Ion Trap Tandem Mass Spectrometry." *Analytica Chimica Acta* 670 (1–2): 63–71. https://doi.org/10.1016/j.aca.2010. 04.023.

European Monitoring Centre for Drugs and Drug Addiction. 2018. "European Drug Report:Trends and Developments 2018." 2018. http://www.emcdda.europa.eu/publications/edr/trends-developments/ 2018_en.

———. 2019. "New Psychoactive Substances (NPS)." 2019. http://www.emcdda.europa.eu/topics/nps_en.

European Monitoring Centre for Drugs and Drug Adiction. 2019. "European Drug Report 2019: Trends and Developments." https://doi.org/http:// www.emcdda.europa.eu/system/files/publications/11364/20191724_T DAT19001ENN_PDF.pdf.

Fehn, Josef, and Gerhard Megges. 1985. "Detection of O6-Monoacetylmorphine in Urine Samples by GC/MS as Evidence for Heroin Use." *Journal of Analytical Toxicology* 9 (3): 134–38.

Feliu, Catherine, Aurélie Fouley, Hervé Millart, Claire Gozalo, Hélène Marty, and Zoubir Djerada. 2015. "Toxicologie Clinique et Analytique Des Opiacés, de La Cocaïne et Des Amphétamines." *Annales de Biologie Clinique* 73 (1): 54–69. https://doi.org/10.1684/abc.2014. 1009.

Feng, Yanru, Min Zheng, Xue Zhang, Kai Kang, Weijun Kang, Kaoqi Lian, and Jie Yang. 2019. "Analysis of Four Antidepressants in Plasma and Urine by Gas Chromatography-Mass Spectrometry Combined with Sensitive and Selective Derivatization." *Journal of Chromatography A* 1600 (August): 33–40. https://doi.org/10.1016/J.CHROMA.2019. 04.038.

Fernández, Nicolás, Laura Marina Cabanillas, Nancy Mónica Olivera, and Patricia Noemí Quiroga. 2019. "Optimization and Validation of Simultaneous Analyses of Ecgonine, Cocaine, and Seven Metabolites in Human Urine by Gas Chromatography–mass Spectrometry Using a One-Step Solid-Phase Extraction." *Drug Testing and Analysis* 11 (2): 361–73. https://doi.org/10.1002/dta.2547.

Fiehn, Oliver. 2008. "Extending the Breadth of Metabolite Profiling by Gas Chromatography Coupled to Mass Spectrometry." *TrAC Trends in Analytical Chemistry* 27 (3): 261–69.

Fonseca, B.M. Da, I.E.D. Moreno, M. Barroso, S. Costa, J.A. Queiroz, and E. Gallardo. 2013. "Determination of Seven Selected Antipsychotic Drugs in Human Plasma Using Microextraction in Packed Sorbent and Gas Chromatography-Tandem Mass Spectrometry." *Analytical and Bioanalytical Chemistry* 405 (12). https://doi.org/10.1007/s00216-012-6695-y.

Gallardo, E., M. Barroso, C. Margalho, A. Cruz, D. N. Vieira, and M. López-Rivadulla. 2006a. "Determination of Parathion in Biological Fluids by Means of Direct Solid-Phase Microextraction." *Analytical and Bioanalytical Chemistry* 386 (6): 1717–26. https://doi.org/10.1007/s00216-006-0735-4.

———. 2006b. "Determination of Quinalphos in Blood and Urine by Direct Solid-Phase Microextraction Combined with Gas Chromatography-Mass Spectrometry." *Journal of Chromatography B: Analytical Technologies in the Biomedical and Life Sciences* 832 (1): 162–68. https://doi.org/10.1016/j.jchromb.2005.12.029.

———. 2006c. "Solid-Phase Microextraction for Gas Chromatographic/ Mass Spectrometric Analysis of Dimethoate in Human Biological Samples." *Rapid Communications in Mass Spectrometry* 20 (5): 865–69. https://doi.org/10.1002/rcm.2388.

Gallardo, E., M. Barroso, and J. A. Queiroz. 2009a. "Current Technologies and Considerations for Drug Bioanalysis in Oral Fluid." *Bioanalysis*. Future Science Ltd. https://doi.org/10.4155/bio.09.23.

Gallardo, E., M. Barroso, and J.A. Queiroz. 2009b. "LC-MS: A Powerful Tool in Workplace Drug Testing." *Drug Testing and Analysis* 1 (3): 109–15. https://doi.org/10.1002/dta.26.

Gallardo, E., and J. A. Queiroz. 2008. "The Role of Alternative Specimens in Toxicological Analysis." *Biomedical Chromatography*. https://doi.org/10.1002/bmc.1009.

Garcia, Antonia, and Coral Barbas. 2011. "Gas Chromatography-Mass Spectrometry (GC-MS)-Based Metabolomics." In *Metabolic Profiling*, 191–204. Springer.

Gasse, Angela, Heidi Pfeiffer, Helga Köhler, and Jennifer Schürenkamp. 2016. "Development and Validation of a Solid-Phase Extraction Method Using Anion Exchange Sorbent for the Analysis of Cannabinoids in Plasma and Serum by Gas Chromatography-Mass Spectrometry." *International Journal of Legal Medicine* 130 (4): 967–74. https://doi.org/10.1007/s00414-016-1368-6.

———. 2018. "8β-OH-THC and 8β,11-DiOH-THC-Minor Metabolites with Major Informative Value?" *International Journal of Legal Medicine* 132 (1): 157–64. https://doi.org/10.1007/s00414-017-1692-5.

Gentili, Stefano, Claudia Mortali, Luisa Mastrobattista, Paolo Berretta, and Simona Zaami. 2016. "Determination of Different Recreational Drugs in Sweat by Headspace Solid-Phase Microextraction Gas Chromatography Mass Spectrometry (HS-SPME GC/MS): Application to Drugged Drivers." *Journal of Pharmaceutical and Biomedical Analysis* 129: 282–87. https://doi.org/10.1016/j.jpba.2016.07.018.

Gérardin, André, Françoise Abadie, and Joëlle Laffont. 1975. "GLC Determination of Carbamazepine Suitable for Pharmacokinetic Studies." *Journal of Pharmaceutical Sciences* 64 (12): 1940–42.

Gervais, Joel R., and Gregory A. Hobbs. 2016. "Use of an Acetyl Derivative to Improve GC-MS Determination of Norbuprenorphine in the Presence of High Concentrations of Buprenorphine in Urine." *Journal of Analytical Toxicology* 40 (3): 208–12. https://doi.org/10.1093/jat/bkw001.

Goldberg, Mark A., Joseph Gal, Arthur K. Cho, and Donald J. Jenden. 1979. "Metabolism of Dimethoxymethyl Phenobarbital (Eterobarb) in Patients with Epilepsy." *Annals of Neurology: Official Journal of the American Neurological Association and the Child Neurology Society* 5 (2): 121–26.

Goldberger, Bruce A., Yale H. Caplan, Tom Maguire, and Edward J. Cone. 1991. "Testing Human Hair for Drugs of Abuse. III. Identification of Heroin and 6-Acetylmorphine as Indicators of Heroin Use." *Journal of Analytical Toxicology* 15 (5): 226–31.

Gonçalves, Joana, Tiago Rosado, Sofia Soares, Ana Y. Simão, Débora Caramelo, Ângelo Luís, Nicolás Fernández, Mário Barroso, Eugenia Gallardo, and Ana Paula Duarte. 2019. "Cannabis and Its Secondary Metabolites: Their Use as Therapeutic Drugs, Toxicological Aspects, and Analytical Determination." *Medicines* 6 (31). https://doi.org/10.3390/medicines6010031.

Górecki, Tadeusz, James Harynuk, and Ognjen Panić. 2004. "The Evolution of Comprehensive Two-dimensional Gas Chromatography (GC× GC)." *Journal of Separation Science* 27 (5-6): 359–79.

Grinstead, Gregory F. 1991. "A Closer Look at Acetyl and Pentafluoropropionyl Derivatives for Quantitative Analysis of Morphine and Codeine by Gas Chromatography/Mass Spectrometry." *Journal of Analytical Toxicology* 15 (6): 293–98. https://doi.org/10.1093/jat/15.6.293.

Grundmann, Milan, Ivana Kacirova, and Romana Urinovska. 2014. "Therapeutic Drug Monitoring of Atypical Antipsychotic Drugs." *Acta Pharmaceutica* 64 (4): 387–401. https://doi.org/10.2478/acph-2014-0036.

Gupta, Ram N., Francis Eng, and Mohan L. Gupta. 1979. "Gas-Chromatographic Analysis for Valproic Acid as Phenacyl Esters." *Clinical Chemistry* 25 (7): 1303–5.

Guthery, Bill, Tom Bassindale, Alan Bassindale, Colin T. Pillinger, and Geraint H. Morgan. 2010. "Qualitative Drug Analysis of Hair Extracts by Comprehensive Two-Dimensional Gas Chromatography/Time-of-

Flight Mass Spectrometry." *Journal of Chromatography A* 1217 (26): 4402–10. https://doi.org/10.1016/j.chroma.2010.04.020.

Gyllenhaal, Olle, and Agneta Albinsson. 1978. "Gas Chromatographic Determination of Valproate in Minute Serum Samples after Extractive Methylation." *Journal of Chromatography A* 161: 343–46.

Hallbach, Jürgen, Hermann Vogel, and Walter G. Guder. 1997. "Determination of Lamotrigine, Carbamazepine and Carbamazepine Epoxide in Human Serum by Gas Chromatography Mass Spectrometry." *Clinical Chemistry and Laboratory Medicine* 35 (10): 755–60.

Harris, D.C. 2007. *Quantitative Chemical Analysis*. 7th ed. New York: W. H. Freeman and Company.

Hasegawa, Chika, Takeshi Kumazawa, Seisaku Uchigasaki, Xiao-Pen Lee, Keizo Sato, Masaru Terada, and Kunihiko Kurosaki. 2011. "Determination of Dextromethorphan in Human Plasma Using Pipette Tip Solid-Phase Extraction and Gas Chromatography–mass Spectrometry." *Analytical and Bioanalytical Chemistry* 401 (7): 2215–23. https://doi.org/10.1007/s00216-011-5324-5.

Hattori, Hideki, Osamu Suzuki, and Hans Brandenberger. 1986. "Positive- and Negative-Ion Mass Spectrometry of Butyrophenones." *Journal of Chromatography B: Biomedical Sciences and Applications* 382: 135–45.

Heinl, Sonja, Oliver Lerch, and Freidoon Erdmann. 2016. "Automated GC–MS Determination of Δ^9-Tetrahydrocannabinol, Cannabinol and Cannabidiol in Hair." *Journal of Analytical Toxicology* 40 (7): 498–503. https://doi.org/10.1093/jat/bkw047.

Hites, Ronald A. 2016. "Development of Gas Chromatographic Mass Spectrometry." *Analytical Chemistry* 88 (14): 6955–61. https://doi.org/10.1021/acs.analchem.6b01628.

Hommes, F. A., J. R. G. Kuipers, J. D. Elema, J. F. Jansen, and J. H. P. Jonxis. 1968. "Propionicacidemia, a New Inborn Error of Metabolism." *Pediatric Research* 2 (6): 519.

Horning, E_C_, and M-G_ Horning. 1971. "Metabolic Profiles: Gas-Phase Methods for Analysis of Metabolites." *Clinical Chemistry* 17 (8): 802–9.

Hosli, R., A. Tobler, S. Konig, and S. Muhlebach. 2013. "A Quantitative Phenytoin GC-MS Method and Its Validation for Samples from Human Ex Situ Brain Microdialysis, Blood and Saliva Using Solid-Phase Extraction." *Journal of Analytical Toxicology* 37 (2): 102–9. https://doi.org/10.1093/jat/bks137.

Huang, Wei, Wilmo Andollo, and William Lee Hearn. 1992. "A Solid Phase Extraction Technique for the Isolation and Identification of Opiates in Urine." *Journal of Analytical Toxicology* 16 (5): 307–10. https://doi.org/10.1093/jat/16.5.307.

Hübschmann, H.-J. 2009. *Handbook of GC/MS: Fundamentals and Applications*. Edited by H.-J. Hübschmann. 2nd ed. Weinheim: WILEY-VCH Verlag GmbH & Co. KGaA.

Hulshoff, Abram, and Hendrik Roseboom. 1979. "Determination of Valproic Acid (DI-N-Propyl Acetic Acid) in Plasma by Gas-Liquid Chromatography with Pre-Column Butylation." *Clinica Chimica Acta* 93 (1): 9–13.

Hurst, Harrell E., David R. Jones, Charles H. Jarboe, and Hans deBree. 1981. "Determination of Clovoxamine Concentration in Human Plasma by Electron Capture Gas Chromatography." *Clinical Chemistry* 27 (7): 1210–12.

Ikeda, Kayo, Kazuro Ikawa, Toshihiro Kozumi, Satoko Yokoshige, Shunji Horikawa, and Norifumi Morikawa. 2012. "Development and Validation of a GC-EI-MS Method with Reduced Adsorption Loss for the Quantification of Olanzapine in Human Plasma." *Analytical and Bioanalytical Chemistry* 403 (7): 1823–30. https://doi.org/10.1007/s00216-012-5802-4.

Ikeda, Kayo, Kazuro Ikawa, Satoko Yokoshige, Satoshi Yoshikawa, and Norifumi Morikawa. 2014. "Gas Chromatography-Electron Ionization-Mass Spectrometry Quantitation of Valproic Acid and Gabapentin, Using Dried Plasma Spots, for Therapeutic Drug Monitoring in in-Home

Medical Care." *Biomedical Chromatography* 28 (12): 1756–62. https://doi.org/10.1002/bmc.3217.

Inturrisi, Charles E., Mitchell B. Max, Kathleen M. Foley, Michael Schultz, Seung-Uon Shin, and Raymond W. Houde. 1984. "The Pharmacokinetics of Heroin in Patients with Chronic Pain." *New England Journal of Medicine* 310 (19): 1213–17.

Isenschmid, Daniel S., Barry S. Levine, and Yale H. Caplan. 1988. "A Method for the Simultaneous Determination of Cocaine, Benzoylecgonine, and Ecgonine Methyl Ester in Blood and Urine Using Gc/Eims with Derivatization to Produce High Mass Molecular Ions." *Journal of Analytical Toxicology* 12 (5): 242–45. https://doi.org/10.1093/jat/12.5.242.

Ishikawa, Aline Akemi, Dayanne Mozaner Bordin, Eduardo Geraldo de Campos, Lucas Blanes, Philip Doble, and Bruno Spinosa De Martinis. 2018. "A Gas Chromatography–Mass Spectrometry Method for Toxicological Analysis of MDA, MDEA and MDMA in Vitreous Humor Samples from Victims of Car Accidents." *Journal of Analytical Toxicology* 42 (9): 661–66. https://doi.org/10.1093/jat/bky044.

Issaq, Haleem J., Que N. Van, Timothy J. Waybright, Gary M. Muschik, and Timothy D. Veenstra. 2009. "Analytical and Statistical Approaches to Metabolomics Research." *Journal of Separation Science* 32 (13): 2183–99.

IUPAC. 2019. "Gold Book, Selected Ion Monitoring in Mass Spectrometry." 2019.

Jain, Naresh C., Thomas C. Sneath, Robert D. Budd, and Wai J. Leung. 1975. "Gas Chromatographic/Thin-Layer Chromatographic Analysis of Acetylated Codeine and Morphine in Urine." *Clinical Chemistry* 21 (10): 1486–89.

Jang, Moonhee, Wonkyung Yang, Sujin Jeong, Sohyung Park, and Jihyun Kim. 2016. "A Fatal Case of Paramethoxyamphetamine Poisoning and Its Detection in Hair." *Forensic Science International* 266: e27–31. https://doi.org/10.1016/j.forsciint.2016.06.030.

Jarque, Pilar, Antonia Roca, Isabel Gomila, Valeria Noce, Bernardino Barcelo, and Julia Klein. 2018. "Quantification of Methamphetamine

«Shabu» in Biological Matrices to Detect Prenatal Exposure: A Case Report and a Literature Review." *Current Pharmaceutical Biotechnology* 19 (2): 163–74. https://doi.org/10.2174/138920101 9666180427111735.

Joya, Xavier, Mitona Pujadas, María Falcón, Ester Civit, Oscar Garcia-Algar, Oriol Vall, Simona Pichini, Aurelio Luna, and Rafael de la Torre. 2010. "Gas Chromatography–mass Spectrometry Assay for the Simultaneous Quantification of Drugs of Abuse in Human Placenta at 12th Week of Gestation." *Forensic Science International* 196 (1–3): 38–42. https://doi.org/10.1016/j.forsciint.2009.12.044.

Juenke, J., and Gwendolyn A. McMillin. 2009. "Analytical Support of Classical Anticonvulsant Drug Monitoring Beyond Immunoassay: Application of Chromatographic Methods." In *Advances in Chromatographic Techniques for Therapeutic Drug Monitoring*, 87–103. CRC Press.

Kanani, Harin, Panagiotis K. Chrysanthopoulos, and Maria I. Klapa. 2008. "Standardizing Gc–ms Metabolomics." *Journal of Chromatography B* 871 (2): 191–201.

Kanani, Harin H., and Maria I. Klapa. 2007. "Data Correction Strategy for Metabolomics Analysis Using Gas Chromatography–mass Spectrometry." *Metabolic Engineering* 9 (1): 39–51.

Kapetanović, Izet M., and Harvey J. Kupferberg. 1980. "Stable Isotope Methodology and Gas Chromatography Mass Spectrometry in a Pharmacokinetic Study of Phenobarbital." *Biomedical Mass Spectrometry* 7 (2): 47–52.

Karila, Laurent, Geneviève Lafaye, Amandine Scocard, Olivier Cottencin, and Amine Benyamina. 2018. "MDPV and α-PVP Use in Humans: The Twisted Sisters." *Neuropharmacology* 134: 65–72. https://doi.org/ 10.1016/j.neuropharm.2017.10.007.

Karschner, Erin L., Allan J. Barnes, Ross H. Lowe, Karl B. Scheidweiler, and Marilyn A. Huestis. 2010. "Validation of a Two-Dimensional Gas Chromatography Mass Spectrometry Method for the Simultaneous Quantification of Cannabidiol, Δ9-Tetrahydrocannabinol (THC), 11-Hydroxy-THC and 11-nor-9-Carboxy-THC in Plasma." *Analytical and*

Bioanalytical Chemistry 397 (2): 603. https://doi.org/10.1007/S00216-010-3599-6.

Kerrigan, Sarah. 2015. "Improved Detection of Synthetic Cathinones in Forensic Toxicology Samples : Thermal Degradation and Analytical Considerations." *U. S. Department of Justice*.

Kieliba, Tobias, Oliver Lerch, Hilke Andresen-Streichert, Markus A. Rothschild, and Justus Beike. 2019. "Simultaneous Quantification of THC-COOH, OH-THC, and Further Cannabinoids in Human Hair by Gas Chromatography-Tandem Mass Spectrometry with Electron Ionization Applying Automated Sample Preparation." *Drug Testing and Analysis* 11 (2): 267–78. https://doi.org/10.1002/dta.2490.

Kim, Jin Young, Jae Chul Cheong, Jae Il Lee, and Moon Kyo In. 2011. "Improved Gas Chromatography-Negative Ion Chemical Ionization Tandem Mass Spectrometric Method for Determination of 11-nor-Δ9-Tetrahydrocannabinol-9-Carboxylic Acid in Hair Using Mechanical Pulverization and Bead-Assisted Liquid-Liquid Extraction." *Forensic Science International* 206 (1–3): e99-102. https://doi.org/10.1016/j.forsciint.2011.01.013.

Kind, Tobias, Gert Wohlgemuth, Do Yup Lee, Yun Lu, Mine Palazoglu, Sevini Shahbaz, and Oliver Fiehn. 2009. "FiehnLib: Mass Spectral and Retention Index Libraries for Metabolomics Based on Quadrupole and Time-of-Flight Gas Chromatography/Mass Spectrometry." *Analytical Chemistry* 81 (24): 10038–48.

Kintz, Pascal, and Patrice Mangin. 1995. "Simultaneous Determination of Opiates, Cocaine and Major Metabolites of Cocaine in Human Hair by Gas Chromotography/Mass Spectrometry (GC/MS)." *Forensic Science International* 73 (2): 93–100. https://doi.org/10.1016/0379-0738(95)01725-X.

Koek, Maud M., Renger H. Jellema, Jan van der Greef, Albert C. Tas, and Thomas Hankemeier. 2011. "Quantitative Metabolomics Based on Gas Chromatography Mass Spectrometry: Status and Perspectives." *Metabolomics* 7 (3): 307–28. https://doi.org/10.1007/s11306-010-0254-3.

Kovatsi, Leda, Konstantinos Rentifis, Dimitrios Giannakis, Samuel Njau, and Victoria Samanidou. 2011. "Disposable Pipette Extraction for Gas Chromatographic Determination of Codeine, Morphine, and 6-Monoacetylmorphine in Vitreous Humor." *Journal of Separation Science* 34 (14): 1716–21. https://doi.org/10.1002/jssc.201100124.

Krasowski, Matthew D. 2010. "Therapeutic Drug Monitoring of the Newer Anti-Epilepsy Medications." *Pharmaceuticals* 3 (6): 1909–35. https://doi.org/10.3390/ph3061909.

Kuhara, Tomiko. 2005. "Gas Chromatographic–mass Spectrometric Urinary Metabolome Analysis to Study Mutations of Inborn Errors of Metabolism." *Mass Spectrometry Reviews* 24 (6): 814–27.

Kupferberg, Harvey J. 1970. "Quantitative Estimation of Diphenylhydantoin, Primidone and Phenobarbital in Plasma by Gas-Liquid Chromatography." *Clinica Chimica Acta* 29 (2): 283–88.

Kuś, Piotr, Joachim Kusz, Maria Książek, Ewelina Pieprzyca, and Marcin Rojkiewicz. 2017. "Spectroscopic Characterization and Crystal Structures of Two Cathinone Derivatives: N-Ethyl-2-Amino-1-Phenylpropan-1-One (Ethcathinone) Hydrochloride and N-Ethyl-2-Amino-1-(4-Chlorophenyl)Propan-1-One (4-CEC) Hydrochloride." *Forensic Toxicology* 35 (1): 114–24. https://doi.org/10.1007/s11419-016-0345-6.

Kyle, P. B. 2017. "Chapter 7 - Toxicology: GCMS." In, edited by Hari Nair and William B. T. - Mass Spectrometry for the Clinical Laboratory Clarke, 131–63. San Diego: Academic Press. https://doi.org/https://doi.org/10.1016/B978-0-12-800871-3.00007-9.

la Torre, Carolina Sánchez de, María A. Martínez, and Elena Almarza. 2005. "Determination of Several Psychiatric Drugs in Whole Blood Using Capillary Gas–liquid Chromatography with Nitrogen Phosphorus Detection: Comparison of Two Solid Phase Extraction Procedures." *Forensic Science International* 155 (2–3): 193–204.

la Torre, Rafael de, Jordi Ortuño, José A. Pascual, Susana González, and Jordi Ballesta. 1998. "Quantitative Determination of Tricyclic Antidepressants and Their Metabolites in Plasma by Solid-Phase Extraction (Bond-Elut TCA) and Separation by Capillary Gas

Chromatography with Nitrogen-Phosphorous Detection." *Therapeutic Drug Monitoring* 20 (3): 340–46.

Lakshmi HimaBindu, M. R., S. Angala Parameswari, and C. Gopinath. 2013. "A Review on GC-MS and Method Development and Validation." *International Journal of Pharmaceutical Quality Assurance* 4 (3): 42–51.

Ledberg, Anders. 2015. "The Interest in Eight New Psychoactive Substances before and after Scheduling." *Drug and Alcohol Dependence* 152: 73–78. https://doi.org/10.1016/j.drugalcdep.2015.04.020.

Lee, H. M., C. W. Lee, G. Lee, H. M. Lee, and C. W. Lee. 1991. "Determination of Morphine and Codeine in Blood and Bile by Gas Chromatography with a Derivatization Procedure." *Journal of Analytical Toxicology* 15 (4): 182–87. https://doi.org/10.1093/jat/15.4.182.

Lee, Xiao-Pen, Chika Hasegawa, Takeshi Kumazawa, Natsuko Shinmen, Yukiko Shoji, Hiroshi Seno, and Keizo Sato. 2008. "Determination of Tricyclic Antidepressants in Human Plasma Using Pipette Tip Solid-phase Extraction and Gas Chromatography–mass Spectrometry." *Journal of Separation Science* 31 (12): 2265–71.

Lehmann, Sabrina, Tobias Kieliba, Justus Beike, Mario Thevis, and Katja Mercer-Chalmers-Bender. 2017. "Determination of 74 New Psychoactive Substances in Serum Using Automated In-Line Solid-Phase Extraction-Liquid Chromatography-Tandem Mass Spectrometry." *Journal of Chromatography B: Analytical Technologies in the Biomedical and Life Sciences* 1064 (April): 124–38. https://doi.org/10.1016/j.jchromb.2017.09.003.

Lensmeyer, Gary L. 1977. "Isothermal Gas Chromatographic Method for the Rapid Determination of Carbamazepine ('Tegretol') as Its TMS Derivative." *Clinical Toxicology* 11 (4): 443–54.

Lerch, Oliver, Oliver Temme, and Thomas Daldrup. 2014. "Comprehensive Automation of the Solid Phase Extraction Gas Chromatographic Mass Spectrometric Analysis (SPE-GC/MS) of Opioids, Cocaine, and Metabolites from Serum and Other Matrices." *Analytical and*

Bioanalytical Chemistry 406 (18): 4443–51. https://doi.org/10.1007/s00216-014-7815-7.

Lindon, John C., Jeremy K. Nicholson, and Elaine Holmes. 2011. *The Handbook of Metabonomics and Metabolomics*. Elsevier.

Liu, Zaiyou, and John B. Phillips. 1991. "Comprehensive Two-Dimensional Gas Chromatography Using an on-Column Thermal Modulator Interface." *Journal of Chromatographic Science* 29 (6): 227–31.

López-Guarnido, O., I. Álvarez, F. Gil, L. Rodrigo, H. C. Cataño, A. M. Bermejo, M. J. Tabernero, A. Pla, and A. F. Hernández. 2013. "Hair Testing for Cocaine and Metabolites by GC/MS: Criteria to Quantitatively Assess Cocaine Use." *Journal of Applied Toxicology* 33 (8): 838–44. https://doi.org/10.1002/jat.2741.

López, P., A. M. Bermejo, M. J. Tabernero, P. Cabarcos, I. Álvarez, and P. Fernández. 2009. "Cocaine and Opiates Use in Pregnancy: Detection of Drugs in Neonatal Meconium and Urine." *Journal of Analytical Toxicology* 33 (7): 351–55. https://doi.org/10.1093/jat/33.7.351.

Lora-Tamayo, C., T. Tena, and G. Tena. 1987. "Concentrations of Free and Conjugated Morphine in Blood in Twenty Cases of Heroin-Related Deaths." *Journal of Chromatography B: Biomedical Sciences and Applications* 422: 267–73.

Lowe, Ross H., Allan J. Barnes, Elin Lehrmann, William J. Freed, Joel E. Kleinman, Thomas M. Hyde, Mary M. Herman, and Marilyn A. Huestis. 2006. "A Validated Positive Chemical Ionization GC/MS Method for the Identification and Quantification of Amphetamine, Opiates, Cocaine, and Metabolites in Human Postmortem Brain." *Journal of Mass Spectrometry* 41 (2): 175–84. https://doi.org/10.1002/jms.975.

Maciów-Głąb, Martyna, Sebastian Rojek, Karol Kula, and Małgorzata Kłys. 2014. "'New Designer Drugs' in Aspects of Forensic Toxicology." *Archives of Forensic Medicine and Criminology* 1 (1): 20–33. https://doi.org/10.5114/amsik.2014.44587.

Madej, Katarzyna, and Paweł Kościelniak. 2008. "Review of Analytical Methods for Identification and Determination of PHEs and Tricyclic Antidepressants." *Critical Reviews in Analytical Chemistry* 38 (2): 50–66. https://doi.org/10.1080/10408340701804343.

Malaca, Sara, Tiago Rosado, José Restolho, Jesus M. Rodilla, Pedro M. M. Rocha, Lúcia Silva, Cláudia Margalho, Mário Barroso, and Eugenia Gallardo. 2019. "Determination of Amphetamine-Type Stimulants in Urine Samples Using Microextraction by Packed Sorbent and Gas Chromatography-Mass Spectrometry." *Journal of Chromatography. B., Analytical Technologies in the Biomedical and Life Sciences* 1120: 41–50. https://doi.org/10.1016/j.jchromb.2019.04.052.

Mandrioli, R., L. Mercolini, M. A. Saracino, and M. A. Raggi. 2012. "Selective Serotonin Reuptake Inhibitors (SSRIs): Therapeutic Drug Monitoring and Pharmacological Interactions." *Current Medicinal Chemistry* 19 (12): 1846–63. https://doi.org/10.2174/092986712800099749.

Mantovani, Cinthia de Carvalho, Jefferson Pereira E. Silva, Guilherme Forster, Rafael Menck de Almeida, Edna Maria de Albuquerque Diniz, and Mauricio Yonamine. 2018. "Simultaneous Accelerated Solvent Extraction and Hydrolysis of 11-nor-Δ9-Tetrahydrocannabinol-9-Carboxylic Acid Glucuronide in Meconium Samples for Gas Chromatography-Mass Spectrometry Analysis." *Journal of Chromatography. B., Analytical Technologies in the Biomedical and Life Sciences* 1074–1075 (February): 1–7. https://doi.org/10.1016/j.jchromb.2018.01.009.

Margalho, C., F. Corte-Real, M. López-Rivadulla, and E. Gallardo. 2016. "Salvia Divinorum: Toxicological Aspects and Analysis in Human Biological Specimens." *Bioanalysis* 8 (13). https://doi.org/10.4155/bio-2016-0067.

Margalho, Cláudia, Alice Castanheira, Francisco Corte Real, Eugenia Gallardo, and Manuel López-Rivadulla. 2016. "Determination of 'New Psychoactive Substances' in Postmortem Matrices Using Microwave Derivatization and Gas Chromatography–mass Spectrometry." *Journal of Chromatography B* 1020: 14–23. https://doi.org/10.1016/j.jchromb.2016.03.001.

Mariotti, Kristiane de Cássia, Roselena S. Schuh, Priscila Ferranti, Rafael S. Ortiz, Daniele Z. Souza, Flavio Pechansky, Pedro E. Froehlich, and Renata P. Limberger. 2014. "Simultaneous Analysis of Amphetamine-

Type Stimulants in Plasma by Solid-Phase Microextraction and Gas Chromatography-Mass Spectrometry." *Journal of Analytical Toxicology* 38 (7): 432–37. https://doi.org/10.1093/jat/bku063.

Markowitz, J. S., and K. S. Patrick. 1995. "Thermal Degradation of Clozapine-N-Oxide to Clozapine during Gas Chromatographic Analysis." *Journal of Chromatography B: Biomedical Sciences and Applications* 668 (1): 171–74.

Martínez, M. A., C. Sánchez De La Torre, and E. Almarza. 2003. "A Comparative Solid-Phase Extraction Study for the Simultaneous Determination of Fluoxetine, Amitriptyline, Nortriptyline, Trimipramine, Maprotiline, Clomipramine, and Trazodone in Whole Blood by Capillary Gas—liquid Chromatography with Nitrogen-Phosphoru." *Journal of Analytical Toxicology* 27 (6): 353–58.

Masumoto, Ken, Yumiko Tashiro, Kumiko Matsumoto, Akiyoshi Yoshida, Masami Hirayama, and Shin'ichi Hayashi. 1986. "Simultaneous Determination of Codeine and Chlorpheniramine in Human Plasma by Capillary Column Gas Chromatography." *Journal of Chromatography B: Biomedical Sciences and Applications* 381: 323–29.

Maurer, Hans H. 2018. "Mass Spectrometry for Research and Application in Therapeutic Drug Monitoring or Clinical and Forensic Toxicology." *Therapeutic Drug Monitoring* 40 (4): 389–93. https://doi.org/10.1097/FTD.0000000000000525.

Maurer, Hans H. 1992. "Systematic Toxicological Analysis of Drugs and Their Metabolites by Gas Chromatography—mass Spectrometry." *Journal of Chromatography B: Biomedical Sciences and Applications* 580 (1–2): 3–41.

Maurer, Hans, and Karl Pfleger. 1984. "Screening Procedure for Detection of Phenothiazine and Analogous Neuroleptics and Their Metabolites in Urine Using a Computerized Gas Chromatographic—mass Spectrometric Technique." *Journal of Chromatography B: Biomedical Sciences and Applications* 306: 125–45.

Mbughuni, Michael M., Paul J. Jannetto, and Loralie J. Langman. 2016. "Mass Spectrometry Applications for Toxicology." *EJIFCC* 27 (4): 272–87. https://www.ncbi.nlm.nih.gov/pubmed/28149262.

McKay, G., K. Hall, J. K. Cooper, E. M. Hawes, and K. K. Midha. 1982. "Gas Chromatographic—mass Spectrometric Procedure for the Quantitation of Chlorpromazine in Plasma and Its Comparison with a New High-Performance Liquid Chromatographic Assay with Electrochemical Detection." *Journal of Chromatography B: Biomedical Sciences and Applications* 232 (2): 275–82.

McMaster, M. C. 2008. *GC/MS A Practical User's Guide. GC/MS: A Practical Use's Guide.*

McMillin, Gwendolyn A., and Matthew D. Krasowski. 2016. "Therapeutic Drug Monitoring of Newer Antiepileptic Drugs." In *Clinical Challenges in Therapeutic Drug Monitoring*, 101–34. Elsevier.

Mecarelli, Oriano, Pietro Li Voti, Stefano Pro, Francesco Saverio Romolo, Maria Rotolo, Patrizia Pulitano, Neri Accornero, and Nicola Vanacore. 2007. "Saliva and Serum Levetiracetam Concentrations in Patients With Epilepsy." *Therapeutic Drug Monitoring* 29 (3): 313–18. https://doi.org/10.1097/FTD.0b013e3180683d55.

Melent'Ev, A. B. 2004. "Gas Chromatography-Mass Spectrometry Determination of Morphine and Codeine in Blood as Their Propionic Esters." *Journal of Analytical Chemistry* 59 (6): 566–70. https://doi.org/10.1023/B:JANC.0000030880.72908.85.

Mercieca, Gilbert, Sara Odoardi, Marisa Cassar, and Sabina Strano Rossi. 2018. "Rapid and Simple Procedure for the Determination of Cathinones, Amphetamine-like Stimulants and Other New Psychoactive Substances in Blood and Urine by GC–MS." *Journal of Pharmaceutical and Biomedical Analysis* 149: 494–501. https://doi.org/10.1016/j.jpba.2017.11.024.

Meyer, Markus R., Jessica Welter, Armin A. Weber, and Hans H. Maurer. 2011. "Development, Validation, and Application of a Fast and Simple GC–MS Method for Determination of Some Therapeutic Drugs Relevant in Emergency Toxicology." *Therapeutic Drug Monitoring* 33 (5): 649–53. https://doi.org/10.1097/FTD.0b013e3182305409.

Millner, S. N., and C. A. Taber. 1979. "Rapid Gas Chromatographic Determination of Carbamazepine for Routine Therapeutic Monitoring."

Journal of Chromatography B: Biomedical Sciences and Applications 163 (1): 96–102.

Milman, Garry, Allan J. Barnes, Ross H. Lowe, and Marilyn A. Huestis. 2010. "Simultaneous Quantification of Cannabinoids and Metabolites in Oral Fluid by Two-Dimensional Gas Chromatography Mass Spectrometry." *Journal of Chromatography. A* 1217 (9): 1513–21. https://doi.org/10.1016/j.chroma.2009.12.053.

Minoli, M., I. Angeli, A. Ravelli, F. Gigli, and F. Lodi. 2012. "Detection and Quantification of 11-nor-Δ9-Tetrahydrocannabinol-9-Carboxylic Acid in Hair by GC/MS/MS in Negative Chemical Ionization Mode (NCI) with a Simple and Rapid Liquid/Liquid Extraction." *Forensic Science International* 218 (1–3): 49–52. https://doi.org/10.1016/j.forsciint.2011.10.014.

Mohamed, Khaled. 2017. "One-Step Derivatization-Extraction Method for Rapid Analysis of Eleven Amphetamines and Cathinones in Oral Fluid by GC–MS." *Journal of Analytical Toxicology* 41 (7): 639–45. https://doi.org/10.1093/jat/bkx046.

Mohamed, Khaled M., and Abdulsallam Bakdash. 2017. "Comparison of 3 Derivatization Methods for the Analysis of Amphetamine-Related Drugs in Oral Fluid by Gas Chromatography-Mass Spectrometry." *Analytical Chemistry Insights* 12: 1–16. https://doi.org/10.1177/1177390117727.

Möller, M. R., P. Fey, and S. Rimbach. 1992. "Identification and Quantitation of Cocaine and Its Metabolites, Benzoylecgonine and Ecgonine Methyl Ester, in Hair of Bolivian Coca Chewers by Gas Chromatography/Mass Spectrometry." *Journal of Analytical Toxicology* 16 (5): 291–96. https://doi.org/10.1093/jat/16.5.291.

Mondello, Luigi, Peter Quinto Tranchida, Paola Dugo, and Giovanni Dugo. 2008. "Comprehensive Two-dimensional Gas Chromatography-mass Spectrometry: A Review." *Mass Spectrometry Reviews* 27 (2): 101–24.

Morita, Yoshio, Tsuen Ih Ruo, Min Long Lee, and Jr A. J. Atkinson. 1981. "On-Column Propylation Method for Measuring Plasma Valproate

Concentration by Gas Chromatography." *Therapeutic Drug Monitoring* 3 (2): 193–99.

Murray, Kermit K., Robert K. Boyd, Marcos N. Eberlin, G. John Langley, Liang Li, and Yasuhide Naito. 2013. "Definitions of Terms Relating to Mass Spectrometry (IUPAC Recommendations 2013)." *Pure and Applied Chemistry* 85 (7): 1515–1609. https://doi.org/10.1351/PAC-REC-06-04-06.

Nahar, Lutfun, Mingquan Guo, and Satyajit D. Sarker. 2019. "Gas Chromatographic Analysis of Naturally Occurring Cannabinoids: A Review of Literature Published during the Past Decade." *Phytochemical Analysis : PCA*, August, pca.2886. https://doi.org/10.1002/pca.2886.

Nakamoto, Akihiro, Manami Nishida, Takeshi Saito, Izumi Kishiyama, Shota Miyazaki, Katsunori Murakami, Masataka Nagao, and Akira Namura. 2010. "Monolithic Silica Spin Column Extraction and Simultaneous Derivatization of Amphetamines and 3,4-Methylenedioxyamphetamines in Human Urine for Gas Chromatographic-Mass Spectrometric Detection." *Analytica Chimica Acta* 661 (1): 42–46. https://doi.org/10.1016/j.aca.2009.12.013.

Namera, Akira, Maho Kawamura, Akihiro Nakamoto, Takeshi Saito, and Masataka Nagao. 2015. "Comprehensive Review of the Detection Methods for Synthetic Cannabinoids and Cathinones." *Forensic Toxicology* 33 (2): 175–94. https://doi.org/10.1007/s11419-015-0270-0.

Nau, H., W. Wittfoht, H. Schäfer, C. Jakobs, D. Rating, and H. Helge. 1981. "Valproic Acid and Several Metabolites: Quantitative Determination in Serum, Urine, Breast Milk and Tissues by Gas Chromatography—mass Spectrometry Using Selected Ion Monitoring." *Journal of Chromatography B: Biomedical Sciences and Applications* 226 (1): 69–78.

Nelson, Michael H., Angela K. Birnbaum, Peter J. Nyhus, and Rory P. Remmel. 1998. "A Capillary GC-MS Method for Analysis of Phenytoin and [13C3]-Phenytoin from Plasma Obtained from Pulse Dose Pharmacokinetic Studies1." *Journal of Pharmaceutical and Biomedical Analysis* 17 (8): 1311–23.

Nicholson, Jeremy K., John Connelly, John C. Lindon, and Elaine Holmes. 2002. "Metabonomics: A Platform for Studying Drug Toxicity and Gene Function." *Nature Reviews Drug Discovery* 1 (2): 153.

Nikolaou, Panagiota, Ioannis Papoutsis, Artemisia Dona, Chara Spiliopoulou, and Sotiris Athanaselis. 2015. "Development and Validation of a GC/MS Method for the Simultaneous Determination of Levetiracetam and Lamotrigine in Whole Blood." *Journal of Pharmaceutical and Biomedical Analysis* 102 (January): 25–32. https://doi.org/10.1016/j.jpba.2014.08.034.

Nikolaou, Panagiota, Ioannis Papoutsis, Chara Spiliopoulou, Constantinos Voudris, and Sotiris Athanaselis. 2015. "A Fully Validated Method for the Determination of Lacosamide in Human Plasma Using Gas Chromatography with Mass Spectrometry: Application for Therapeutic Drug Monitoring." *Journal of Separation Science* 38 (2): 260–66. https://doi.org/10.1002/jssc.201400858.

Nishioka, Ryota, Matsumi Takeuchi, Satoshi Kawai, Mikio Nakamura, and Kazuko Kondo. 1985. "Improved Direct Injection Method and Extractive Methylation Method for Determination of Valproic Acid in Serum by Gas Chromatography." *Journal of Chromatography B: Biomedical Sciences and Applications* 342: 89–96.

Niu, Zongliang, Weiwei Zhang, Chunwei Yu, Jun Zhang, and Yingying Wen. 2018. "Recent Advances in Biological Sample Preparation Methods Coupled with Chromatography, Spectrometry and Electrochemistry Analysis Techniques." *TrAC Trends in Analytical Chemistry* 102: 123–46. https://doi.org/https://doi.org/10.1016/j.trac.2018.02.005.

Noorizadeh, Hadi, and Mehrab Noorizadeh. 2012. "QSRR-Based Estimation of the Retention Time of Opiate and Sedative Drugs by Comprehensive Two-Dimensional Gas Chromatography." *Medicinal Chemistry Research* 21 (8): 1997–2005. https://doi.org/10.1007/s00044-011-9727-9.

Norman, Trevor R., and Kay P. Maguire. 1985. "Analysis of Tricyclic Antidepressant Drugs in Plasma and Serum by Chromatographic

Techniques." *Journal of Chromatography B: Biomedical Sciences and Applications* 340: 173–97.

Nowatzke, William, Jianbo Zeng, Al Saunders, Alan Bohrer, John Koenig, and John Turk. 1999. "Distinction among Eight Opiate Drugs in Urine by Gas Chromatography-Mass Spectrometry." *Journal of Pharmaceutical and Biomedical Analysis* 20 (5): 815–28. https://doi.org/10.1016/S0731-7085(99)00086-2.

Odoardi, Sara, Francesco Saverio Romolo, and Sabina Strano-Rossi. 2016. "A Snapshot on NPS in Italy: Distribution of Drugs in Seized Materials Analysed in an Italian Forensic Laboratory in the Period 2013-2015." *Forensic Science International* 265: 116–20. https://doi.org/10.1016/j.forsciint.2016.01.037.

Ottaviani, Giovanni, Roberto Cameriere, Marta Cippitelli, Rino Froldi, Giovanna Tassoni, Massimiliano Zampi, and Mariano Cingolani. 2017. "Determination of Drugs of Abuse in a Single Sample of Human Teeth by a Gas Chromatography-Mass Spectrometry Method." *Journal of Analytical Toxicology* 41 (1): 32–36. https://doi.org/10.1093/jat/bkw105.

Papoutsis, Ioannis, Alaa Khraiwesh, Panagiota Nikolaou, Constantinos Pistos, Chara Spiliopoulou, and Sotirios Athanaselis. 2012. "A Fully Validated Method for the Simultaneous Determination of 11 Antidepressant Drugs in Whole Blood by Gas Chromatography–mass Spectrometry." *Journal of Pharmaceutical and Biomedical Analysis* 70 (November): 557–62. https://doi.org/10.1016/j.jpba.2012.05.007.

Pasikanti, Kishore K., P. C. Ho, and E. C. Y. Chan. 2008. "Gas Chromatography/Mass Spectrometry in Metabolic Profiling of Biological Fluids." *Journal of Chromatography B* 871 (2): 202–11.

Patteet, Lisbeth, Delphine Cappelle, Kristof E. Maudens, Cleo L. Crunelle, Bernard Sabbe, and Hugo Neels. 2015. "Advances in Detection of Antipsychotics in Biological Matrices." *Clinica Chimica Acta* 441 (Supplement C): 11–22. https://doi.org/https://doi.org/10.1016/j.cca.2014.12.008.

Pego, A. M.F., F. L. Roveri, R. Y. Kuninari, V. Leyton, I. D. Miziara, and M. Yonamine. 2017. "Determination of Cocaine and Its Derivatives in

Hair Samples by Liquid Phase Microextraction (LPME) and Gas Chromatography–mass Spectrometry (GC–MS)." *Forensic Science International* 274 (May): 83–90. https://doi.org/10.1016/j.forsciint.2016.12.024.

Pelição, Fabrício Souza, Mariana Dadalto Peres, Jauber Fornaciari Pissinate, and Bruno Spinosa De Martinis. 2014. "A One-Step Extraction Procedure for the Screening of Cocaine, Amphetamines and Cannabinoids in Postmortem Blood Samples." *Journal of Analytical Toxicology* 38 (6): 341–48. https://doi.org/10.1093/jat/bku039.

Pellegrini, Manuela, Adriana Casá, Emilia Marchei, Roberta Pacifici, Ruth Mayné, Vanessa Barbero, Oscar Garcia-Algar, and Simona Pichini. 2006. "Development and Validation of a Gas Chromatography-Mass Spectrometry Assay for Opiates and Cocaine in Human Teeth." *Journal of Pharmaceutical and Biomedical Analysis* 40 (3): 662–68. https://doi.org/10.1016/j.jpba.2005.07.003.

Perchalski, Robert J., and B. J. Wilder. 1974. "Rapid Gas—Liquid Chromatographic Determination of Carbamazepine in Plasma." *Clinical Chemistry* 20 (4): 492–93.

Peres, Mariana Dadalto, Fabrício Souza Pelição, Bruno Caleffi, and Bruno Spinosa De Martinis. 2014. "Simultaneous Quantification of Cocaine, Amphetamines, Opiates and Cannabinoids in Vitreous Humor." *Journal of Analytical Toxicology* 38 (1): 39–45. https://doi.org/10.1093/jat/bkt093.

Petersen, Erling N., Erik Bechgaard, Rodney J. Sortwell, and Lennart Wetterberg. 1978. "Potent Depletion of 5HT from Monkey Whole Bloob by a New 5HT Uptake Inhibitor, Paroxetine (FG 7051)." *European Journal of Pharmacology* 52 (1): 115–19.

Pfleger, Karl, Hans H. Maurer, and Armin Weber. 1992. *Mass Spectral and GC Data of Drugs, Poisons, Pesticides, Pollutants and Their Metabolites. Part 1: Methods, Tables, Indexes.* VCH Verlagsgesellschaft mbH.

Pichini, Simona, Roberta Pacifici, Ilaria Altieri, Manuela Pellegrini, and Piergiorgio Zuccaro. 1999. "Determination of Opiates and Cocaine in Hair as Trimethylsilyl Derivatives Using Gas Chromatography-Tandem

Mass Spectrometry." *Journal of Analytical Toxicology* 23 (5): 343–48. https://doi.org/10.1093/jat/23.5.343.

Pommier, F., A. Sioufi, and J. Godbillon. 1997. "Simultaneous Determination of Imipramine and Its Metabolite Desipramine in Human Plasma by Capillary Gas Chromatography with Mass-Selective Detection." *Journal of Chromatography B: Biomedical Sciences and Applications* 703 (1–2): 147–58.

Pragst, Fritz. 2004. "*Pitfalls in Hair Analysis.*" https://gtfch.org/cms/images/stories/media/tk/tk71_2/Pragst1.pdf.

Prata, Margarida, Andreia Ribeiro, David Figueirinha, Tiago Rosado, David Oppolzer, José Restolho, André R.T.S. Araújo, Suzel Costa, Mário Barroso, and Eugenia Gallardo. 2019. "Determination of Opiates in Whole Blood Using Microextraction by Packed Sorbent and Gas Chromatography-Tandem Mass Spectrometry." *Journal of Chromatography A* 1602: 1–10. https://doi.org/10.1016/j.chroma.2019.05.021.

Prata, Vanessa de M., Elissandro S. Emídio, and Haroldo S. Dorea. 2012. "New Catalytic Ultrasound Method for Derivatization of 11-nor-Δ9-Tetrahydrocannabinol-9-Carboxylic Acid in Urine, with Analysis by GC-MS/MS." *Analytical and Bioanalytical Chemistry* 403 (2): 625–32. https://doi.org/10.1007/s00216-012-5827-8.

Pujadas, Mitona, Simona Pichini, Ester Civit, Elena Santamariña, Katherine Perez, and Rafael de la Torre. 2007. "A Simple and Reliable Procedure for the Determination of Psychoactive Drugs in Oral Fluid by Gas Chromatography-Mass Spectrometry." *Journal of Pharmaceu-tical and Biomedical Analysis* 44 (2): 594–601. https://doi.org/10.1016/j.jpba.2007.02.022.

Purschke, Kirsten, Sonja Heinl, Oliver Lerch, Freidoon Erdmann, and Florian Veit. 2016. "Development and Validation of an Automated Liquid-Liquid Extraction GC/MS Method for the Determination of THC, 11-OH-THC, and Free THC-Carboxylic Acid (THC-COOH) from Blood Serum." *Analytical and Bioanalytical Chemistry* 408 (16): 4379–88. https://doi.org/10.1007/s00216-016-9537-5.

Racamonde, Inés, Eugenia Villaverde-de-Sáa, Rosario Rodil, José Benito Quintana, and Rafael Cela. 2012. "Determination of Δ9-Tetrahydro-cannabinol and 11-nor-9-Carboxy-Δ9-Tetrahydro-cannabinol in Water Samples by Solid-Phase Microextra-ction with on-Fiber Derivatization and Gas Chromatography–mass Spectrometry." *Journal of Chromatography A* 1245 (July): 167–74. https://doi.org/10.1016/j.chroma.2012.05.017.

Raharjo, Tri J., and Robert Verpoorte. 2004. "Methods for the Analysis of Cannabinoids in Biological Materials: A Review." *Phytochemical Analysis* 15 (2): 79–94. https://doi.org/10.1002/pca.753.

Rana, Sumandeep, Rakesh K. Garg, and Anu Singla. 2014. "Rapid Analysis of Urinary Opiates Using Fast Gas Chromatography-Mass Spectrometry and Hydrogen as a Carrier Gas." *Egyptian Journal of Forensic Sciences* 4 (3): 100–107. https://doi.org/10.1016/j.ejfs.2014.03.001.

Rani, Susheela, and Ashok Kumar Malik. 2012. "A Novel Microextraction by Packed Sorbent-Gas Chromatography Procedure for the Simultaneous Analysis of Antiepileptic Drugs in Human Plasma and Urine." *Journal of Separation Science* 35 (21): 2970–77. https://doi.org/10.1002/jssc.201200439.

Rege, Arvind B., Juan J. L. Lertora, Luann E. White, and William J. George. 1984. "Rapid Analysis of Valproic Acid by Gas Chromatography." *Journal of Chromatography B: Biomedical Sciences and Applications* 309: 397–402.

Restek. 2019. "Why Derivatize?" 2019. https://www.restek.com/pdfs/CFTS1269.pdf.

Rhoden, Liliane, Marina Venzon Antunes, Paulina Hidalgo, Cleber Álvares da Silva, and Rafael Linden. 2014. "Simple Procedure for Determination of Valproic Acid in Dried Blood Spots by Gas Chromatography–mass Spectrometry." *Journal of Pharmaceutical and Biomedical Analysis* 96 (August): 207–12. https://doi.org/10.1016/j.jpba.2014.03.044.

Ripple, M. G., B. A. Goldberger, Y. H. Caplan, M. G. Blitzer, and S. Schwartz. 1992. "Detection of Cocaine and Its Metabolites in Human Amniotic Fluid." *Journal of Analytical Toxicology* 16 (5): 328–31. https://doi.org/10.1093/jat/16.5.328.

Ritz, Dennis P., C. Gerald Warren, and Daniel T. Teitelbaum. 1975. "Single Extraction GLC Analysis of Six Commonly Prescribed Antiepileptic Drugs." *Clinical Toxicology* 8 (3): 311–24.

Romolo, F. S., M. C. Rotolo, I. Palmi, R. Pacifici, and A. Lopez. 2003. "Optimized Conditions for Simultaneous Determination of Opiates, Cocaine and Benzoylecgonine in Hair Samples by GC-MS." *Forensic Science International* 138 (1–3): 17–26. https://doi.org/10.1016/j.forsciint.2003.07.013.

Rosado, T., M. Barroso, D. N. Vieira, and E. Gallardo. 2019. "Determination of Selected Opiates in Hair Samples Using Microextraction by Packed Sorbent: A New Approach for Sample Clean-Up." *Journal of Analytical Toxicology* 43 (6): 465–76. https://doi.org/10.1093/jat/bkz029.

Rosado, T., L. Fernandes, M. Barroso, and E. Gallardo. 2017. "Sensitive Determination of THC and Main Metabolites in Human Plasma by Means of Microextraction in Packed Sorbent and Gas Chromatography-Tandem Mass Spectrometry." *Journal of Chroma-tography. B, Analytical Technologies in the Biomedical and Life Sciences* 1043 (February): 63–73. https://doi.org/10.1016/j.jchromb.2016.09.007.

Rosado, Tiago, Alexandra Gonçalves, Cláudia Margalho, Mário Barroso, and Eugenia Gallardo. 2017. "Rapid Analysis of Cocaine and Metabolites in Urine Using Microextraction in Packed Sorbent and GC/MS." *Analytical and Bioanalytical Chemistry* 409 (8): 2051–63. https://doi.org/10.1007/s00216-016-0152-2.

Rosado, Tiago, Alexandra Gonçalves, Ana Martinho, Gilberto Alves, Ana Paula Duarte, Fernanda Domingues, Samuel Silvestre, et al. 2017. "Simultaneous Quantification of Antidepressants and Metabolites in Urine and Plasma Samples by GC–MS for Therapeutic Drug Monitoring." *Chromatographia* 80 (2): 301–28. https://doi.org/10.1007/s10337-017-3240-3.

Rosado, Tiago, Joana Gonçalves, Ângelo Luís, Sara Malaca, Sofia Soares, Duarte Nuno Vieira, Mário Barroso, and Eugenia Gallardo. 2018. "Synthetic Cannabinoids in Biological Specimens: A Review of Current Analytical Methods and Sample Preparation Techniques." *Bioanalysis* 10 (19): 1609–23. https://doi.org/10.4155/bio-2018-0150.

Rosado, Tiago, David Oppolzer, Belinda Cruz, Mário Barroso, Samira Varela, Victor Oliveira, Carlos Leitão, and Eugenia Gallardo. 2018. "Development and Validation of a Gas Chromatography/Tandem Mass Spectrometry Method for Simultaneous Quantitation of Several Antipsychotics in Human Plasma and Oral Fluid." *Rapid Communications in Mass Spectrometry* 32 (23): 2081–95. https://doi.org/10.1002/rcm.8087.

Salgado-Petinal, Carmen, J. Pablo Lamas, Carmen Garcia-Jares, Maria Llompart, and Rafael Cela. 2005. "Rapid Screening of Selective Serotonin Re-Uptake Inhibitors in Urine Samples Using Solid-Phase Microextraction Gas Chromatography–mass Spectrometry." *Analytical and Bioanalytical Chemistry* 382 (6): 1351–59.

Salomone, Alberto, Joseph J. Palamar, Enrico Gerace, Daniele Di Corcia, and Marco Vincenti. 2017. "Hair Testing for Drugs of Abuse and New Psychoactive Substances in a High-Risk Population." *Journal of Analytical Toxicology* 41 (5): 376–81. https://doi.org/10.1093/jat/bkx020.

Santos, Rosimeire Resende Dos, Maria José Nunes Paiva, Júlio César Veloso, Philippe Serp, Zenilda De Lourdes Cardeal, and Helvécio Costa Menezes. 2017. "Efficient Extraction Method Using Magnetic Carbon Nanotubes to Analyze Cocaine and Benzoylecgonine in Breast Milk by GC/MS." *Bioanalysis* 9 (21): 1655–66. https://doi.org/10.4155/bio-2017-0140.

Sawada, Yuji, Kenji Akiyama, Akane Sakata, Ayuko Kuwahara, Hitomi Otsuki, Tetsuya Sakurai, Kazuki Saito, and Masami Yokota Hirai. 2008. "Widely Targeted Metabolomics Based on Large-Scale MS/MS Data for Elucidating Metabolite Accumulation Patterns in Plants." *Plant and Cell Physiology* 50 (1): 37–47.

Schwertner, Harvey A., Horace E. Hamilton, and Jack E. Wallace. 1978. "Analysis for Carbamazepine in Serum by Electron-Capture Gas Chromatography." *Clinical Chemistry* 24 (6): 895–99.

Segura, Jordi, Cristiana Stramesi, Alicia Redón, Montserrat Ventura, Carlos J. Sanchez, Gerard González, Luis San, and Maria Montagna. 1999. "Immunological Screening of Drugs of Abuse and Gas

Chromatographic-Mass Spectrometric Confirmation of Opiates and Cocaine in Hair." *Journal of Chromatography B: Biomedical Sciences and Applications* 724 (1): 9–21. https://doi.org/10.1016/S0378-4347(98)00531-3.

Serafin, Michelle C., Kasandra M. Paulemon, Zachary J. Fuller, and William E. Bronner. 2017. "Modified Method for Detection of Benzoylecgonine in Human Urine by GC-MS: Derivatization Using Pentafluoropropanol/Acetic Anhydride." *Journal of Analytical Toxicology* 41 (4): 340–41. https://doi.org/10.1093/jat/bkx007.

Shah, Iltaf, Bayan Al-Dabbagh, Alaa Eldin Salem, Saber A. A. Hamid, Neak Muhammad, and Declan P. Naughton. 2019. "A Review of Bioanalytical Techniques for Evaluation of Cannabis (Marijuana, Weed, Hashish) in Human Hair." *BMC Chemistry* 13 (1): 106. https://doi.org/10.1186/s13065-019-0627-2.

Shen, Min, Ping Xiang, Hejian Wu, Baohua Shen, and Zhongjie Huang. 2002. "Detection of Antidepressant and Antipsychotic Drugs in Human Hair." *Forensic Science International* 126 (2): 153–61.

Shulaev, Vladimir. 2006. "Metabolomics Technology and Bioinformatics." *Briefings in Bioinformatics* 7 (2): 128–39.

Sigma-Aldrich. 2019. "Mixed-Mode SPE Improves Extraction of Pharmaceutical Compounds from Biological Fluids." 2019.

Smith, Michael L., Daniel C. Nichols, Paula Underwood, Zachary Fuller, Matthew A. Moser, Charles LoDico, David A. Gorelick, Matthew N. Newmeyer, Marta Concheiro, and Marilyn A. Huestis. 2014. "Morphine and Codeine Concentrations in Human Urine Following Controlled Poppy Seeds Administration of Known Opiate Content." *Forensic Science International* 241 (August): 87–90. https://doi.org/10.1016/j.forsciint.2014.04.042.

Snozek, Christine L. H., Loralie J. Langman, and Steven W. Cotten. 2019. "An Introduction to Drug Testing: The Expanding Role of Mass Spectrometry." In *Methods in Molecular Biology (Clifton, N.J.)*, 1872:1–10. https://doi.org/10.1007/978-1-4939-8823-5_1.

Soares, Sofia, Tiago Rosado, Mário Barroso, Duarte Nuno Vieira, and Eugenia Gallardo. 2019. "Organophosphorus Pesticide Determination in Biological Specimens: Bioanalytical and Toxicological Aspects." *International Journal of Legal Medicine* 133 (6): 1763–84. https://doi.org/10.1007/s00414-019-02119-9.

Society of Hair Testing. 2019. "Recommendations for Hair Testing in Forensic Cases Society of Hair Testing Criteria for Mass Spectrometric Analysis." 2019. http://www.soht.org/pdf/Consensus_on_Hair_Analysis.pdf.

Sporkert, F., and F. Pragst. 2000. "Use of Headspace Solid-Phase Microextraction (HS-SPME) in Hair Analysis for Organic Compounds." *Forensic Science International* 107 (1–3): 129–48. https://doi.org/10.1016/S0379-0738(99)00158-9.

Stefanelli, Fabio, Federica Giorgia Pesci, Mario Giusiani, and Silvio Chericoni. 2018. "A Novel Fast Method for Aqueous Derivatization of THC, OH-THC and THC-COOH in Human Whole Blood and Urine Samples for Routine Forensic Analyses." *Biomedical Chromatography : BMC* 32 (4): e4136. https://doi.org/10.1002/bmc.4136.

Stout, Peter R., Jay M. Gehlhausen, Carl K. Horn, and Kevin L. Klette. 2002. "Evaluation of a Solid-Phase Extraction Method for Benzoylecgonine Urine Analysis in a High-Throughput Forensic Urine Drug-Testing Laboratory." *Journal of Analytical Toxicology* 26 (7): 401–5. https://doi.org/10.1093/jat/26.7.401.

Strong, John M., Thomas Abe, Erlch L. Gibbs, and Arthur J. Atkinson. 1974. "Plasma Levels of Methsuximide and N-desmethylmethsuxi-mide during Methsuximide Therapy." *Neurology* 24 (3): 250.

Suzuki, O., H. Hattori, M. Asano, and H. Brandenberger. 1986. "Positive and Negative Ion Mass Spectrometry of Tricyclic Antidepressants." *Zeitschrift Für Rechtsmedizin* 97 (4): 239–50.

Tabarra, Inês, Sofia Soares, Tiago Rosado, Joana Gonçalves, Ângelo Luís, Sara Malaca, Mário Barroso, Thomas Keller, José Restolho, and Eugenia Gallardo. 2019. "Novel Synthetic Opioids - Toxicological Aspects and Analysis." *Forensic Sciences Research* 4 (2): 111–40. https://doi.org/10.1080/20961790.2019.1588933.

Tanaka, K., M. A. Budd, M. L. Efron, and K. J. Isselbacher. 1966. "Isovaleric Acidemia: A New Genetic Defect of Leucine Metabolism." *Proceedings of the National Academy of Sciences of the United States of America* 56 (1): 236.

Thompson, John A., and Sanford P. Markey. 1975. "Quantitative Metabolic Profiling of Urinary Organic Acids by Gas Chromatography-Mass Spectrometry. Comparison of Isolation Methods." *Analytical Chemistry* 47 (8): 1313–21.

Thurman, E.M., and M.S. Mills. 1998. *Solid-Phase Extraction: Principles and Practice*. 1st ed. New York: John Wiley & Sons, Inc.

Torok-Both, George A., Glen B. Baker, Ronald T. Coutts, Kevin F. McKenna, and Launa J. Aspeslet. 1992. "Simultaneous Determination of Fluoxetine and Norfluoxetine Enantiomers in Biological Samples by Gas Chromatography with Electron-Capture Detection." *Journal of Chromatography B: Biomedical Sciences and Applications* 579 (1): 99–106.

Tosoni, S., C. Signorini, and A. Albertini. 1983. "Gas-Chromatographic Determination of Valproic Acid in Serum without Derivatization." *Clinical Chemistry* 29 (5): 990.

Tremaine, Larry M., and Evelyn A. Joerg. 1989. "Automated Gas Chromatographic—electron-Capture Assay for the Selective Serotonin Uptake Blocker Sertraline." *Journal of Chromatography B: Biomedical Sciences and Applications* 496: 423–29.

Tupper, Nancy L., Elizabeth B. Solow, and Caroline P. Kenfield. 1978. "A Method for Esterification of Valproic Acid for Gas-Liquid Chromatography: Clinical Data from Epileptic Patients." *Journal of Analytical Toxicology* 2 (5): 203–6.

Uddin, Mohammad N., Victoria F. Samanidou, and Ioannis N. Papadoyannis. 2011. "Bio-Sample Preparation and Analytical Methods for the Determination of Tricyclic Antidepressants." *Bioanalysis* 3 (1): 97–118. https://doi.org/10.4155/bio.10.160.

Ulrich, S., and J. Martens. 1997. "Solid-Phase Microextraction with Capillary Gas-Liquid Chromatography and Nitrogen-Phosphorus Selective Detection for the Assay of Antidepressant Drugs in Human

Plasma." *Journal of Chromatography B: Biomedical Sciences and Applications* 696 (2): 217–34.

Unceta, Nora, M. Aranzazu Goicolea, and Ramón J. Barrio. 2011. "Analytical Procedures for the Determination of the Selective Serotonin Reuptake Inhibitor Antidepressant Citalopram and Its Metabolites." *Biomedical Chromatography* 25 (1): 238–57. https://doi.org/10.1002/bmc.1542.

United Nations Office on Drugs and Crime. 2018. "Global Smart Update Volumne 19: Understanding the Synthetic Drug Market: The NPS Factor," 1–12. https://www.unodc.org/documents/scientific/Global_Smart_Update_2018_Vol.19.pdf.

Valente, Maria João, Paula Guedes De Pinho, Maria De Lourdes Bastos, Félix Carvalho, and Márcia Carvalho. 2014. "Khat and Synthetic Cathinones: A Review." *Archives of Toxicology* 88 (1): 15–45. https://doi.org/10.1007/s00204-013-1163-9.

Van, A. Langenhove, J. E. Biller, K. Biemann, and T. R. Browne. 1982. "Simultaneous Determination of Phenobarbital and P-Hydroxyphenobarbital and Their Stable Isotope Labeled Analogs by Gas Chromatography Mass Spectrometry." *Biomedical Mass Spectrometry* 9 (5): 201–7.

Vandemark, Frank L., and Reginald F. Adams. 1976. "Ultramicro Gas-Chromatographic Analysis for Anticonvulsants, with Use of a Nitrogen-Selective Detector." *Clinical Chemistry* 22 (7): 1062–65.

Viant, Mark R. 2008. "Recent Developments in Environmental Metabolomics." *Molecular Biosystems* 4 (10): 980–86.

Vu-Duc, Trinh, and André Vernay. 1990. "Simultaneous Detection and Quantitation of O6-monoacetylmorphine, Morphine and Codeine in Urine by Gas Chromatography with Nitrogen Specific and/or Flame Ionization Detection." *Biomedical Chromatography* 4 (2): 65–69. https://doi.org/10.1002/bmc.1130040206.

Wasels, Richard, and Francine Belleville. 1994. "Gas Chromatographic-Mass Spectrometric Procedures Used for the Identification and Determination of Morphine, Codeine and 6-Monoacetylmorphine."

Journal of Chromatography A 674 (1–2): 225–34. https://doi.org/10.1016/0021-9673(94)85227-8.

Watelle, Mirjam, Paul Demedts, Francis Franck, Peter Paul De Deyn, Annick Wauters, and Hugo Neels. 1997. "Analysis of the Antiepileptic Phenyltriazine Compound Lamotrigine Using Gas Chromatography with Nitrogen Phosphorus Detection." *Therapeutic Drug Monitoring* 19 (4): 460–64.

Way, Barbara A., Douglas Stickle, Mary E. Mitchell, John W. Koenig, and John Turk. 1998. "Isotope Dilution Gas Chromatographic-Mass Spectrometric Measurement of Tricyclic Antidepressant Drugs. Utility of the 4-Carbethoxyhexafluorobutyryl Derivatives of Secondary Amines." *Journal of Analytical Toxicology* 22 (5): 374–82.

Welthagen, Werner, Robert A. Shellie, Joachim Spranger, Michael Ristow, Ralf Zimmermann, and Oliver Fiehn. 2005. "Comprehensive Two-Dimensional Gas Chromatography–time-of-Flight Mass Spectrometry (GC× GC-TOF) for High Resolution Metabolomics: Biomarker Discovery on Spleen Tissue Extracts of Obese NZO Compared to Lean C57BL/6 Mice." *Metabolomics* 1 (1): 65–73.

Wiergowski, Marek, Justyna Aszyk, Michał Kaliszan, Kamila Wilczewska, Jacek Sein Anand, Agata Kot-Wasik, and Zbigniew Jankowski. 2017. "Identification of Novel Psychoactive Substances 25B-NBOMe and 4-CMC in Biological Material Using HPLC-Q-TOF-MS and Their Quantification in Blood Using UPLC–MS/MS in Case of Severe Intoxications." *Journal of Chromatography B* 1041–1042: 1–10. https://doi.org/https://doi.org/10.1016/j.jchromb.2016.12.018.

Wikoff, William R., Jon A. Gangoiti, Bruce A. Barshop, and GARY SIuzDAK. 2007. "Metabolomics Identifies Perturbations in Human Disorders of Propionate Metabolism." *Clinical Chemistry* 53 (12): 2169–76.

Wilkins, Diana G., Douglas E. Rollins, Jared Seaman, Heather Haughey, Gerald Krueger, and Rodger Foltz. 1995. "Quantitative Determination of Codeine and Its Major Metabolites in Human Hair by Gas Chromatography-Positive Ion Chemical Ionization Mass Spectrometry:

A Clinical Application." *Journal of Analytical Toxicology* 19 (5): 269–74. https://doi.org/10.1093/jat/19.5.269.

Wille, Sarah M. R., Paul Van Hee, Hugo M. Neels, Carlos H. Van Peteghem, and Willy E. Lambert. 2007. "Comparison of Electron and Chemical Ionization Modes by Validation of a Quantitative Gas Chromatographic–mass Spectrometric Assay of New Generation Antidepressants and Their Active Metabolites in Plasma." *Journal of Chromatography A* 1176 (1–2): 236–45.

Willox, S., and Susan E. Foote. 1978. "Simple Method for Measuring Valproate (Epilim) in Biological Fluids." *Journal of Chromatography A* 151 (1): 67–70.

Wohler, A. Shannon, and Alphonse Poklis. 1997. "A Simple, Rapid Gas-Liquid Chromatographic Procedure for the Determination of Valproic Acid in Serum." *Journal of Analytical Toxicology* 21 (4): 306–9.

World Health Organization. 2016. "Drug Use and Road Safety." 2016. https://www.who.int/substance_abuse/drug_use_road_safety/en/.

Wu, Hao, Ruyi Xue, Ling Dong, Taotao Liu, Chunhui Deng, Huazong Zeng, and Xizhong Shen. 2009. "Metabolomic Profiling of Human Urine in Hepatocellular Carcinoma Patients Using Gas Chromatography/Mass Spectrometry." *Analytica Chimica Acta* 648 (1): 98–104. https://doi.org/https://doi.org/10.1016/j.aca.2009.06.033.

Wu, Hao, Ruyi Xue, Zhaoqing Tang, Chunhui Deng, Taotao Liu, Huazong Zeng, Yihong Sun, and Xizhong Shen. 2010. "Metabolomic Investigation of Gastric Cancer Tissue Using Gas Chromatography/Mass Spectrometry." *Analytical and Bioanalytical Chemistry* 396 (4): 1385–95.

Wu, Ya Hsueh, Keh liang Lin, Su Chin Chen, and Yan Zin Chang. 2008. "Integration of GC/EI-MS and GC/NCI-MS for Simultaneous Quantitative Determination of Opiates, Amphetamines, MDMA, Ketamine, and Metabolites in Human Hair." *Journal of Chromatography B: Analytical Technologies in the Biomedical and Life Sciences* 870 (2): 192–202. https://doi.org/10.1016/j.jchromb.2008.06.017.

Xiong, Jun, Jie Chen, Man He, and Bin Hu. 2010. "Simultaneous Quantification of Amphetamines, Caffeine and Ketamine in Urine by Hollow Fiber Liquid Phase Microextraction Combined with Gas Chromatography-Flame Ionization Detector." *Talanta* 82 (3): 969–75. https://doi.org/10.1016/j.talanta.2010.06.001.

Xu, Fengguo, Li Zou, and Choon Nam Ong. 2009. "Multiorigination of Chromatographic Peaks in Derivatized GC/MS Metabolomics: A Confounder That Influences Metabolic Pathway Interpretation." *Journal of Proteome Research* 8 (12): 5657–65.

Xue, Ruyi, Zhenxin Lin, Chunhui Deng, Ling Dong, Taotao Liu, Jiyao Wang, and Xizhong Shen. 2008. "A Serum Metabolomic Investigation on Hepatocellular Carcinoma Patients by Chemical Derivatization Followed by Gas Chromatography/Mass Spectrometry." *Rapid Communications in Mass Spectrometry: An International Journal Devoted to the Rapid Dissemination of Up-to-the-Minute Research in Mass Spectrometry* 22 (19): 3061–68.

Yeh, S. Y. 1973. "Separation and Identification of Morphine and Its Metabolites and Congeners." *Journal of Pharmaceutical Sciences* 62 (11): 1827–29.

Yu, Hsiu-Ying, and Mei-Chin Shih. 1996. "Determination of Valproic Acid in Human Plasma by Capillary Gas Chromatography." *Therapeutic Drug Monitoring* 18 (1): 107–8.

Zeeuw, R. A. De. 1992. "Gas Chromatographic Retention Indices of Toxicologically Relevant Substances and Their Metabolites." *Report of the DFG Commission for Clinical Toxicological Analysis, Special Issue of the TIAFT Bulletin.*

Zeng, Jingbin, Jing Zou, Xinhong Song, Jinmei Chen, Jiaojiao Ji, Bo Wang, Yiru Wang, Jaeho Ha, and Xi Chen. 2011. "A New Strategy for Basic Drug Extraction in Aqueous Medium Using Electrochemically Enhanced Solid-Phase Microextraction." *Journal of Chromatography. A* 1218 (2): 191–96. https://doi.org/10.1016/j.chroma.2010.11.020.

Zhang, Guodong, Alvin V. Terry Jr, and Michael G. Bartlett. 2008. "Bioanalytical Methods for the Determination of Antipsychotic Drugs." *Biomedical Chromatography* 22 (7): 671–87.

Zuba, Dariusz. 2012. "Identification of Cathinones and Other Active Components of 'legal Highs' by Mass Spectrometric Methods." *Trends in Analytical Chemistry* 32: 15–30. https://doi.org/10.1016/j.trac.2011.09.009.

Zucchella, Alessandra, Cristiana Stramesi, Lucia Politi, Luca Morini, and Aldo Polettini. 2007. "Treatments against Hair Loss May Hinder Cocaine and Metabolites Detection." *Forensic Science, Medicine, and Pathology* 3 (June): 93–100. https://doi.org/10.1007/s12024-007-0011-8.

In: Gas Chromatography
Editor: Percy Henrichon

ISBN: 978-1-53617-350-5
© 2020 Nova Science Publishers, Inc.

Chapter 2

APPLICATION OF SENSOR GAS CHROMATOGRAPHY IN FORENSIC MEDICINE

Hiroshi Kinoshita[1,*], Naoko Tanaka[1], Mostofa Jamal[1], Asuka Ito[1], Mitsuru Kumihashi[1], Tadayoshi Yamashita[1], Shoji Kimura[1], Yasuhiko Kimura[1], Kunihiko Tsutsui[2], Shuji Matsubara[3] and Kiyoshi Ameno[1]

[1]Department of Forensic Medicine, Faculty of Medicine, Kagawa University, Miki, Kita, Kagawa, Japan
[2]Health Sciences, Faculty of Medicine, Kagawa University, Miki, Kita, Kagawa, Japan

---

[*] Corresponding Author's E-mail: kinochin@med.kagawa-u.ac.jp.

³Postgraduate Clinical Education Center,
Kagawa University Hospital, Miki,
Kita, Kagawa, Japan

ABSTRACT

Gas chromatography is widely used for toxicological analyses in forensics for various kinds of gaseous or volatile substances, such as ethanol, carbon monoxide, hydrogen, cyanide. The present study investigated the analysis of these substances using sensor gas chromatography equipped with a semiconductor gas sensor detector. The simplicity, ease of handling, and high sensitivity of this method allow results to be obtained rapidly, which may provide valuable information for forensic diagnosis.

Keywords: sensor gas chromatography, gaseous or volatile substances, ethanol, carbon monoxide

1. INTRODUCTION

Gas chromatography (GC) is widely used for toxicological analyses [1-7]. A wide range of gaseous and volatile compounds can be identified and quantitated using GC. The GC system consists of a gas supply, column, detector, and data processor. The separation of compounds occurs in the column under a continuous stream of carrier gas flow. The carrier gas is an inert gas, such as nitrogen, hydrogen, or helium, supplied from a cylinder with a regulator.

The signal from the detector is processed by computer and converted into a chromatogram. Component identification is accomplished by the detector, based on peak retention time. Because the signal from the detector is usually proportional to the amount of each component across a concentration range, quantification can often be done from peak height or peak area. Thus, GC provides valuable information for forensic analysis.

Sensor gas chromatography (sGC) is used in breath analyzers and other industrial devices [8-13]. The present report describes the application of sGC to forensic medicine.

2. PRINCIPLE OF sGC

The sGC is a gas chromatograph equipped with a semiconductor sensor as a detector [8-13]. The sGC system consists of four components: the gas supply, a column in a thermostatic oven, the detector, and a data processor (Figure 1).

Ambient air that passes through a filter is supplied by a pump as a carrier gas. Therefore, this apparatus does not require a gas cylinder. The column for sGC was constructed from a Teflon tube packed with separation material with the gas sensor directly connected to the outlet of the column. The semiconductor gas sensor uses a compound such as stannic oxide as the detector, which senses reducing or oxidizing gases in air from a change in electric resistance [14]. This detector is highly selective toward the target gases and very sensitive. The computer both controls the sGC system and performs the data analysis.

3. FORENSIC APPLICATION OF sGC

A portable sGC (Figure 2) requires only about 2 - 5 mL gas sample for a single measurement. The static head-space (HS) technique is used to prepare liquid samples. This technique is simple and can minimize artifacts during analysis [15].

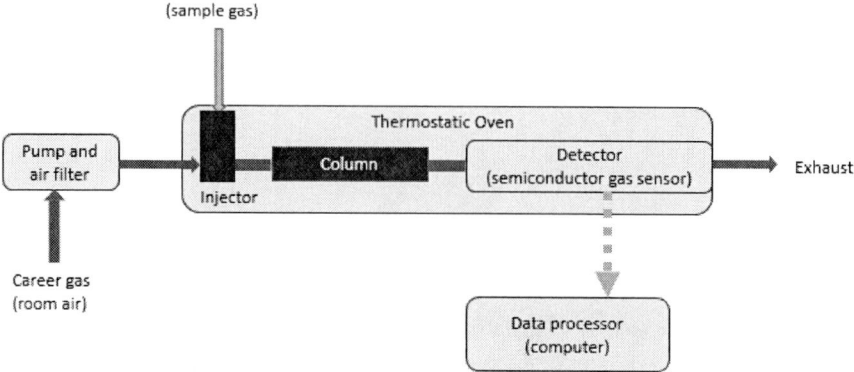

Figure 1. Basic configuration of sensor gas chromatography.

Figure 2. The sensor gas chromatography system (ODSA, FIS Inc, Itami, Japan).

Due to its many advantages, such as easy handling, simple construction, and portability, the sGC has been used for the analysis of various gases and volatile substances in medical practice [8-13].

3.1. Carbon monoxide (CO)

The CO is a poisonous gas [16, 17] that is the leading cause of poisoning deaths in Japan [18, 19]. Therefore, postmortem investigations for CO poisoning are important in forensic practice. CO is produced mainly by incomplete combustion of fuel or other carbon-containing compounds, such as wood fires, automobile exhaust, or faulty heating equipment [16, 20].

The CO binds to hemoglobin and form carboxyhemoglobin (CO-Hb), which disrupts oxygen transport and causes hypoxia [17, 20]. Clinical symptoms roughly correlate with CO-Hb levels [20]. Headache, dizziness, nausea, and weakness are observed at levels of 10 – 30% CO-Hb; a level of CO-Hb that exceeds 60% becomes life-threatening. The level of CO-Hb is important for the diagnosis of CO poisoning or fire-related death [16, 17, 20].

The blood level of CO-Hb can be measured several ways [21] such as spectrophotometry [22, 23], GC with a thermal conductivity detector (TCD) [24-26], and oximetry [27, 28].

For GC measurement, the CO is released from hemoglobin upon treatment with 85% phosphoric acid [25] or potassium ferricyanide in a sealed HS vial [26]. The sGC, which has high sensitivity for CO [limit of quantitation (LOQ) = 1μg/mL], is applicable to CO poisoning cases, with concentrated phosphoric acid (85%) used as the CO liberating agent during sample preparation [29].

Since endogenous CO involves regulatory roles *in vivo*, according to a recent study [30], sGC can be used as a biological monitor to determine CO concentration in breath [12].

3.2. Cyanide

Hydrogen cyanide and its sodium and potassium salts are potent toxic substances [16]. Hydrogen cyanide also is present in smoke from nitrogen-containing plastic fires [16].

Poisoning occurs by inhalation of hydrogen cyanide gas or by ingestion of a cyanide salt. Because cyanide is absorbed quickly from the lungs and stomach, the symptoms of acute poisoning, such as headache, tachypnea, dizziness, coma, and seizures, can occur within 10 – 20 minutes in severe cases [20] due to cytochrome oxidase inhibition and impairment of cellular respiration [16].

Determination of the cyanide level in blood is important for forensic diagnosis [20]. A GC with a nitrogen phosphorus detector (NPD) or flame thermo-ionic detector (FTD) can be used for quantitative analysis [31, 32]. A fatal concentration of cyanide in blood is approximately 3 - 5µg/mL [33].

To quantitate cyanide using GC, it is extracted from a sample in the HS vial by addition of concentrated phosphoric acid or sulfuric acid, followed by introduction into the GC [31, 32]. sGC has good sensitivity toward cyanide (LOQ = 0.05µg/mL) and can be used for the diagnosis of cyanide poisoning [34].

3.3. Ethanol and Acetaldehyde

As alcoholic beverages are ingested by people in many cultures, ethanol is an important drug to test for in forensic practice [35]. Quantitation of ethanol is performed not only on postmortem samples, but also on blood and breath samples to identify people driving under the influence, or people who may have experienced drug-facilitated sexual assault, or accidental or intentional poisoning [17].

Blood, urine, and breath samples can be used for quantitative ethanol analysis. The infrared analyzer is widely employed for ethanol breath measurements [35]. However, sGC can also be used to measure the amounts of ethanol and acetaldehyde in breath [13]. Acetaldehyde, a primary metabolite of ethanol, is also detectable by sGC [13]. However, acetaldehyde concentration in breath may also reflect the genetic polymorphism of aldehyde dehydrogenase 2 [13].

3.4. Hydrogen

Hydrogen gas (H_2) is odorless, colorless, and tasteless and acts as an antioxidant. Its use as a therapeutic agent in cerebral infarction, cardiac ischemia, and organ transplantation has been described in recent reports [36-38]. The H_2 concentration in breath can be used in clinical practice to diagnose gastrointestinal diseases, such as bacterial overgrowth in the small intestine or carbohydrate malabsorption (*e.g.*, lactose intolerance) [39, 40].

Because H_2 is generated by bacterial fermentation of carbohydrates, it may be used as a marker of putrefaction [41]. sGC has high sensitivity for H_2 (LOQ = 1 μg/mL) and can be used to measure H_2 in forensic practice [42].

3.5. Hydrogen Sulfide (H_2S)

The H_2S is a colorless, flammable gas with a rotten egg odor [16, 20]. It is formed in volcanic gases and hot springs, and during the putrefaction of some organic substances. Since a small amount of H_2S is produced in oral cavity, it can cause oral malodor [9], which can be quantitated using sGC [9].

The H_2S is highly toxic and causes cellular asphyxia by inhibition of cytochrome oxidase. Symptoms such as headache, nausea, vomiting, dizziness, loss of consciousness, and respiratory failure can occur after H_2S exposure [20].

A blood concentration of sulfide greater than 0.08 μg/mL can lead to death [43]. Since GC with a flame photometric detector (FPD) is used to quantitate H_2S [44, 45], sGC can also be employed for H_2S analysis in forensic practice.

3.6. Toluene, Ethylbenzene and Xylene Isomers

Toluene is an aromatic hydrocarbon used in industrial and household articles, paint, paint thinner, and glues [16]. It is one of the most abused solvents [46]. Xylene and ethylbenzene are used as a solvent for pesticides [16, 47-54].

These volatile substances are easily detectable by HS-GC. Detecting one of these solvents may be a good indicator of pesticide ingestion [52-54]. The sGC instruments for detection of toluene, ethylbenzene, and xylene are available commercially and may be applicable for pesticide ingestion screening.

4. CONCLUSION AND FUTURE PERSPECTIVES

The use of sGC in medical practice has been applied as a breath analyzer [9-13] and to detect gaseous or volatile substances in forensic medicine [29, 34, 42]. The data obtained by sGC may be useful to determining the cause of death in some cases and provide valuable information for forensic analysis. Further application in forensic practice is expected.

CONFLICTS OF INTEREST

The authors declare that they have no conflicts of interest regarding the publication of this paper.

ACKNOWLEDGMENT

This work was supported by JSPS KAKENHI Grant-in-Aid for Scientific Research (C) Number 15K08873.

REFERENCES

[1] Nishikawa, M., Nishioka, H., Tsuchihashi, H. (2000). New analytical techniques for determination of toxins -4- Gas chromatograph. *Chudoku Kenkyu*, 13: 191 - 199.

[2] Suzuki, O. (2002). Detection methods. In: Suzuki, O., Watanabe, K. editors. Drugs and poisons in humans, *a handbook of practical analysis*. pp. 33 - 43. Berlin, Heidelberg: Springer-Verlag.

[3] Dawling, S. (2004). Gas chromatography. In: Moffat, A. C., Osselton, M. D., Widdop, B., editors. *Clarke's analysis of drugs and poisons in pharmaceuticals, body fluids and postmortem materials* (3rd Ed). pp. 425 - 499. London, Pharmaceutical Press.

[4] Jackson, A. R. W., Jackson, J. M. (2004). *Forensic Science*. Essex: Pearson Education Limited.

[5] Uchiyama, K., Komori, K. (2012). Expert Series for Analytical Chemistry Instrumentation Analysis: Vol. 7. *Gas chromatography*. Tokyo, Kyoritsu.

[6] Stuart, B. H. (2013). *Forensic analytical techniques*. West Sussex, John Wiley & Sons, Ltd.

[7] Takeuchi, T. (2016). *Gas chromatography*. In: Imura, H., Hinoue, T. editors. Instrumental methods in analytical chemistry. pp. 178 - 193. Kyoto: Kagaku Dojin.

[8] Ohta, K., Terai, H., Kimura, I., Tanaka, K. (1999). Simultaneous determination of hydrogen, methane and carbon monoxide in water by gas chromatography with semiconductor detector. *Anal. Chem.*, 71: 2697 - 2699.

[9] Hanada, M., Koda, H., Onaga, K., Tanaka, K., Okabayashi, T., Itoh, T., Miyazaki, H. (2003). Portable oral malodor analyzer using highly sensitive In_2O_3 gas sensor combined with a simple gas chromatography system. *Anal. Chim. Acta,* 475: 27 - 35.

[10] Sakamoto, M. (2005). Development of portable acetone analyzer for human breath analysis. *J. Japan Association on Odor Environment,* 36: 266 - 269.

[11] Hanada, M., Tanaka, K. (2008). Application to the measuring instrument of a semiconductor gas sensor. *Materials Integration,* 21: 152 - 158.

[12] Ueda, H. (2006). Analysis of carbon monoxide in biological specimen and clinical significance. *J. Japan Association on Odor Environment,* 37: 94 - 98.

[13] Himemiya-Hakucho, A., Yamaji, S., Liu, J., Fujimiya, T. (2018). Relationship between genetic polymorphisms of ALDH2 and breath acetaldehyde level under non-drinking state. *Alcohol and Biomedical Research,* 37: 63 - 64.

[14] Yamazoe, N., Shimanoe, K. (2011). Basic approach to the transducer function of oxide semiconductor gas sensors. *Sensors and Actuators B: Chemical,* 160: 1352 - 1362.

[15] Seto, Y. (1994). Determination of volatile substances in biological samples by headspace gas chromatography. *J. Chromatogr. A,* 674: 25 - 62.

[16] Baselt, R. C. (2008). *Disposition of toxic drugs and chemicals in man.* 8th ed. Foster City: Biochemical Publications.

[17] Powers, R. H., Dean, D. E. (2016). *Forensic Toxicology mechanism and pathology.* Boca raton: CRC Press.

[18] Ministry of Health, Labour and Walfare. *Vital Statistics of Japan.* 1998 - 2015.

[19] Ito, T., Nakamura, Y. (2010). Death from carbon monoxide poisoning in Japan between 1968 - 2007 through data from vital statistics. *Journal of Japanese Society for Emergency Medicine,* 13: 275 - 282.

[20] Elenhorn, M. J., Barceloux, D. G. (1988). *Medical Toxicology diagnosis and treatment of human poisoning.* New York: Elsevier.

[21] Akane, A., Fukui, Y. (1985). A review on the development of carboxyhemoglobin analysis. *Res. Pract. Forens. Med.,* 28: 185 - 190.

[22] Kozuka, H., Niwase, T., Taniguchi, T. (1969). Spectrophotometric determination of carboxyhemoglobin. *Journal of Hygiene Chemistry,* 15: 342 - 345.

[23] Ishizu, H., Ameno, S., Yamamoto, Y., Okamura, Y., Shogano, N. (1982). A simple spectrophotometric method for CO-Hb

determination in blood and its application in legal medicine. *Acta Crim. Japon,* 48: 187 - 196.

[24] Hishida, S., Mizoi, Y. (1976). Studies on the quantitative determination of carbon monoxide in human blood by gas chromatography. *Nihon Hoigaku Zasshi,* 30: 319 - 326.

[25] Cardeal, Z. L., Pradeau, D., Hamon, M., Abdoulaye, I., Pailler, F. M., Lejeune, B. (1993). New calibration method for gas chromatographic assay of carbon monoxide in blood. *J. Anal. Toxicol.,* 17: 193 - 195.

[26] Seto, Y., Kataoka, M., Tsuge, K. (2001). Stability of blood carbon monoxide and hemoglobins during heating. *Forensic Sci. Int.,* 121: 144 - 150.

[27] Tanaka, N., Ameno, K., Jamal, M., Ohkubo, E., Kumihashi, M., Kinoshita, H. Application of oximeter AVOX 4000 for the determination of CO-Hb in the forensic practice. *Res. Pract. Forens. Med.,* 2010; 53: 39 - 43.

[28] Fujihara, J., Kinoshita, H., Tanaka, N., Yasuda, T., Takeshita, H. (2013). Accuracy and usefulness of the AVOXimeter 4000 as routine analysis of carboxyhemoglobin. *J. Forensic Sci.,* 58: 1047 - 1049.

[29] Tanaka, N., Kinoshita, H., Takakura, A., Jamal, M., Kumihashi, M., Uchiyama, Y., Tsutsui, K., Ameno, K. (2014). Application of sensor gas chromatography for the determination of carbon monoxide in forensic medicine. *Current study of Environmental and Medical Sciences*, 7: 9 - 11.

[30] Kajimura, M., Nakanishi, T., Takenouchi, T., Morikawa, T., Hishiki, T., Yukutake, Y., Suematsu, M. (2012). gas biology: tiny molecules controlling metabolic systems. *Respir. Physiol. Neurobiol.,* 184: 139 - 148. Doi: 10.1016/j.resp.2012.03.016.

[31] Yoshida, M., Watabiki, T., Ishida, N. (1989). The quantitative determination of cyanide by FTD-GC. *Nihon Houigaku Zasshi*, 43: 179 - 185.

[32] Seto, Y., Tsunoda, N., Ohta, H., Shinohara, T. (1993). Determination of blood cyanide by headspace gas chromatography with nitrogen-phosphorus detection and using a megabore capillary column. *Analytica Chimica Acta*, 276: 247 - 259.

[33] Seto, Y. (2002). Cyanide. In: Suzuki, O., Watanabe, K., editors. Drugs and poisons in humans, *a handbook of practical analysis*. Berlin: Springer-Verlag, pp. 113 - 122.

[34] Tanaka, N., Hanada, M., Hirano, S., Takakura, A., Jamal, M., Ito, A., Kumihashi, M., Tsutsui, K., Kimura, S., Ameno, K., Ishimaru, I., Kinoshita, H. (2016). Determination of cyanide by sensor gas chromatography, *Japanese Journal of Forensic Pathology*, 22: 79 - 82.

[35] Ferner, R. E. (1996). *Forensic pharmacology medicines, mayhem and malpractice.* Oxford: Oxford university press.

[36] Ohta, S. (2011). Recent progress toward hydrogen medicine: potential of molecular hydrogen for preventive and therapeutic applications. *Curr. Pharm. Des.,* 17: 2241 - 2252.

[37] Zhai, X., Chen, X., Ohta, S., Sun, X. (2014). Review and prospect of the biomedical effects of hydrogen. *Med. Gas Res.,* 4: 19.

[38] Ohsawa, I., Ishikawa, M., Takahashi, K., Watamabe, M., Nishimaki, K., Yamagata, K., Katsura, K., Katayama, Y., Asoh, S., Ohta, S. (2007). Hydrogen acts as a therapeutic antioxidant by selectively reducing cytotoxic oxygen radicals. *Nat. Med.,* 13: 688 - 694.

[39] Ghoshal, U. C. (2011). How to interpret hydrogen breath tests. *J. Neurogastroenterol. Motil.,* 17: 312 - 317.

[40] Rana, S. V., Malik, A. (2014). Hydrogen breath tests in gastrointestinal diseases. *Ind. J. Clin. Biochem.,* 29: 398 - 405.

[41] Bernaldo de Quirós, Y., González-Díaz, O., Møllerløkken, A., Brubakk, A. O., Hjelde, A., Saavedra, P., Fernández, A. (2013). Differentiation at autopsy between in vivo gas embolism and putrefaction using gas composition analysis. *Int. J. Legal Med.,* 127: 437 - 445.

[42] Tanaka, N., Kinoshita, H., Takakura, A., Jamal, M., Kumihashi, M., Ito, A., Tsutsui, K., Kimura, S., Matsubara, S., Ameno, K. (2017). Application of sensor gas chromatography for the determination of hydrogen gas in forensic medicine. *Revista e Mjekësisë Ligjore Shqiptare (Review of Albanian Legal Medicine),* 13: 57 - 63.

[43] Maebashi, K., Iwadate, K., Sakai, K., Takatsu, A., Fukui, K., Aoyagi, M., Ochiai, E., Nagai, T. (2011). Toxicological analysis of 17 autopsy cases of hydrogen sulfide poisoning resulting from the inhalation of intentionally generated hydrogen sulfide gas. *Forensic Sci. Int.*, 207: 91 - 95.

[44] Fukui, Y., Kagawa, M., Takahashi, S., Hata, M., Matsubara, K. (1980). Analysis of sulfur compounds in biological materials by gas chromatography with flame photometric detector. *Nihon Houigaku Zasshi*, 34: 575 - 581.

[45] Tanaka, E., Nakamura, T., Terada, M., Misawa, S., Suzuki, Y., Kuroiwa, Y. (1987). The determination of hydrogen sulfide in fluid and tissue samples by gas chromatography with flame photometric detector. *Eisei Kagaku*, 33: 149 - 152.

[46] Kumamoto, T. (2004). Toluene. *Nihon Rinsho* suppl. 12: 509 - 511.

[47] Yamaguchi, Y., Shirakawa, Y., Ogura, S., Ameno, K., Fuke, C., Ogli, K. (1989). A case of amitraz poisoning. *Jpn. J. Clin. Toxicol.*, 2: 289 - 292.

[48] Yamazaki, M., Terada, M., Kuroki, H., Honda, K., Matoba, R., Mitsukuni, Y. (2001). Pesticide poisoning initially suspected as a natural death. *J. Forensic Sci.*, 46: 165 - 170.

[49] Kinoshita, H., Ameno, K., Ameno, S., Kubota, T., Zhang, X., Ijiri, I., Taniguchi, T., Nishiguchi, M., Ouchi, H., Minami, T., Hishida, S. (2001). A case of burning with fenitrothion ingestion. *Res. Pract. Forens. Med.*, 44: 155 - 159.

[50] Kinoshita, H., Nishiguchi, M., Ouchi, H., Minami, T., Yamamura, T., Yasui, T., Marukawa, S., Ameno, K., Hishida, S. (2005). Methanol: toxicity of the solvent in a commercial product should also be considered. *Hum. Toxicol.*, 24: 1 - 2.

[51] Takayasu, T., Ishida, Y., Nosaka, M., Kawaguchi, M., Kuninaka, Y., Kimura, A., Kondo, T. (2012). High concentration of methidathion detected in a fatal case of organophosphate-poisoning. *Leg. Med.*, 14: 263 - 266.

[52] Kinoshita, H., Tanaka, N., Jamal, M., Kumihashi, M., Tsutsui, K., Ameno, K. (2013). Xylene; a useful marker for agricultural products ingestion. *Soud. Lek.,* 58, 59 - 60.

[53] Tanaka, N., Kinoshita, H., Takakura, A., Jamal, M., Kumihashi, M., Uchiyama, Y., Tsutsui, K., Ameno, K. (2015). Combination of energy-dispersive X-ray fluorescence spectrometry (EDX) and head-space gas chromatography mass spectrometry (HS-GC/MS) is a useful screening tool for stomach contents. *Rom. J. Leg. Med.,* 23, 43 - 44.

[54] Kinoshita, H., Tanaka, N., Takakura, A., Kumihashi, M., Jamal, M., Ito, A., Tsutsui, K., Kimura, S., Nagano, T., Matsubara, S., Ameno, K. (2015). Analysis of stomach contents by head-space gas chromatography/mass spectrometry to screen for ingestion of insecticide. *Revista e Mjekësisë Ligjore Shqiptare,* 11, 85 - 89.

In: Gas Chromatography
Editor: Percy Henrichon

ISBN: 978-1-53617-350-5
© 2020 Nova Science Publishers, Inc.

Chapter 3

TRENDS OF GAS CHROMATOGRAPHY-MASS SPECTROMETRY TECHNIQUES IN FOOD AUTHENTICATION

Oscar Núñez[1,2,3,*]

[1]Department of Chemical Engineering and Analytical Chemistry,
University of Barcelona, Barcelona, Spain
[2]Research Institute in Food Nutrition and Food Safety,
University of Barcelona, Barcelona, Spain
[3]Serra Húnter Fellow, Generalitat de Catalunya, Spain

ABSTRACT

Food adulteration practices are potentially harmful to human health and so food safety and authenticity constitute an important issue in food chemistry. The chemical composition of foodstuffs is an excellent indicator of quality, origin, authenticity and/or adulteration. In general, food adulteration is carried out to increase volume, to mask the presence of inferior quality components, and to replace the authentic substances for the

---

[*] Corresponding Author's Email: oscar.nunez@ub.edu.

seller's economic gain. For instance, a common fraud is the employment of a cheaper similar ingredient, which the consumer has difficulty recognizing and which is difficult to detect by current analytical methodologies. For example, fruit-processed products are common adulterated by addition of water, sugars, fruits of inferior commercial value, secondary extracts of fruits and colors, etc. In other cases, the fraud is related to an incorrect labelling of the food product, for example, when refined or lower quality olive oils are labeled as extra-virgin olive oils. Thus, the development of analytical methodologies to achieve food authentication and to identify food frauds is required.

Gas chromatography coupled to mass spectrometry (GC-MS) is nowadays one of the most employed techniques to address food authenticity issues especially by the non-targeted fingerprinting of the food volatolome (volatile metabolites present in the foodstuffs). Targeted strategies by focusing either in the determination of specific food biomarkers or by the profiling of selected families of chemical compounds such as fatty acids are also employed within GC methodologies for authentication purposes. Compound-specific isotope analysis by isotope ratio mass spectrometry (IRMS) following the on-line combustion of compounds separated by GC has also become a method of choice in the authenticity control of foodstuffs based on the measurement of the isotope distribution at natural abundance level. In addition, multidimensional gas chromatography coupled with mass spectrometry has become also a powerful tool in food analysis, being also employed in food authentication problems. In this chapter, the role of GC-MS techniques for food authentication and the identification and prevention of frauds will be addressed. Coverage of all kind of applications is beyond the scope of the present contribution, so the present chapter will focus on the most relevant applications published in the last years.

Keywords: gas chromatography; mass spectrometry, food authentication, multidimensional gas chromatography, isotope ratio mass spectrometry, fingerprinting; target analysis, non-target analysis

INTRODUCTION

The quality of food products is nowadays a topic of major interest for the society. Initially, people's concerns about foodstuffs focused on the presence of contaminants such as pesticides, veterinary drugs, toxins, food packaging contaminants, etc. Therefore, food safety was paramount for scientists, food producers and society in general. However, in the last decade, people have been considering food safety as a granted issue (all countries need to guarantee the safety of consumed food products by means of regulatory authorities, laboratory control analysis, etc.). Society is then becoming more interested in other value added food attributes related to food quality such as the presence of bioactive substances providing beneficial properties for consumers (functional foods, nutraceuticals, etc.). In fact, many consumers are willing to pay more for food products with certain attributes related to the region of production and origin of that food, the farming and growing practices employed (ecological versus conventional practices), etc., and are becoming more interested in food traceability and authenticity aspects [1]. These aspects have giving rise to the consideration of the protected designations of origin (PDO) of natural foodstuffs as important food quality attributes [2]. Within this context, food integrity and authenticity is also becoming a paramount issue in the food field.

Although international and local regulatory bodies have established important rules concerning the labeling of foodstuffs, it is almost impossible to know and to guarantee the real origin of most of the components of a given food, especially to those that have been processed. In addition, taking into consideration the complexity of the food chain and that many players are involved between production and consumption of a given food product, food manipulation and adulteration practices are raising because it is in fact much easier to conduct food fraud without being easily detected. As an example, Moore et al. observed that olive oil, milk, honey, and saffron were the most common targets for adulteration reported in scholarly journals reviewed between 1980 and 2010 [3]. Generally, food adulteration is carried out to increase the volume, to mask the presence of inferior quality components and to replace the authentic substances for the seller's economic gain. However, it must be considered that the deliberate adulteration of food and its misrepresentation to deceive the final consumers is illegal worldwide. Food adulteration practices not only implies economic consequences, but can also represent an important health issue for consumers when prohibited and potentially toxic substances are added to deceive the organoleptic or nutritional properties of the final adulterated product. That was the case of the addition of melamine during the adulteration of high protein content foods [4], or the addition of the non-permitted artificial pigment called E141 in extra-virgin olive oils [5]. Another important aspect is when accidental exposure to allergenic proteins may result from undeclared allergenic substances by means of food adulteration, fraud or uncontrolled cross-contamination [6].

Considering all this, food adulteration could be potentially harmful to human health and so food safety and quality control constitute an important issue in this field. For this reason, the main players in the food chain, regulatory authorities, food processors, retailers and consumers, are very interested in the certification of food authenticity. Thus, from a commercial and legal point of view, regulatory authorities are requested to continuously update the analytical methods and conditions to validate the authenticity of a certain food product as this may also support law enforcement actions [7]. Consequently, the development of new analytical methodologies to guarantee food integrity and authenticity is required, also considering that food adulteration has become increasingly sophisticated, often being specially designed to avoid detection through routine analytical approaches.

The analysis of food products is a difficult task mainly due to the complexity and the diversity of sample matrices, but also due to the great variability of compounds that can be present, differing in polarity, structures, as well as in concentration levels (from major to trace level compounds). In addition, different aspects need to be considered simultaneously when developing new analytical approaches to guarantee food integrity and authenticity, such as the sample treatment and extraction procedures, the separation and determination approaches, as well as the identification and confirmation strategies [8].

Nowadays, two main analytical approaches are employed when addressing food integrity and authenticity aspects: targeted and non-targeted analysis. Targeted approaches are based on the specific determination of a given group of known selected chemicals, or a group of chemicals belonging to the same family or with similar structural features. The concentration (or the specific signal) of these targeted compounds can then be proposed as food features (markers) of a given product to address its integrity and authenticity. These approaches generally require a previous quantitation step using standards for each targeted analyte, if concentration levels are required. Nevertheless, standards are not always commercially available or are sometimes quite expensive. In addition, the quantitation of some chemicals may also be a difficult task due to the complexity of food matrices, as previously commented, and the possibility of unknown interfering compounds. Regarding targeted approaches where only the instrumental signal of some specific chemicals is employed (without a quantitation step), confirmation strategies are required to guarantee the identity of the targeted chemicals, and their unequivocal identification will always require the use of standards. However, if food markers are perfectly identified and standards are available, targeted analytical methodologies are very appropriate to address food authenticity aspects.

In contrast, non-targeted approaches (based on "metabolomic" fingerprinting) focus on the analysis of instrumental responses without assuming any previous knowledge of relevant or irrelevant food components. These instrumental responses can be, for example, the optical spectroscopic information recorded as a function of time (UV-visible, fluorescence, Near-Infrared spectroscopy, etc.), or the peak intensity values recorded as a function of m/z and retention time when mass spectrometric techniques are employed.

Because of the complexity of food and the great variability of chemical compounds that can be present, the amount of chemical data that can be extracted, especially when dealing with non-targeted fingerprinting approaches, is enormous. As a consequence, in order to extract useful (bio)chemical information from the sample data sets able to allow the characterization, classification and authentication of food products, chemometric data treatments will be required. Among them, multivariate methods such as principal component analysis (PCA) and partial least squares regression-discriminant analysis (PLS-DA) are typically employed for exploratory and classificatory purposes, respectively, in food integrity and authenticity [9].

Chromatographic separation techniques are typically employed when dealing with food fraud and authenticity aspects, and among them, gas chromatography coupled to mass spectrometry (GC-MS) is one of the most employed techniques especially by the non-targeted fingerprinting of the food volatolome (the volatile metabolites present in a given food product). Targeted strategies by focusing either in the determination of specific food biomarkers or by the profiling of selected families of chemical compounds such as fatty acids are also employed within GC methodologies for authentication purposes. In addition, compound-specific isotope analysis by isotope ratio mass spectrometry (IRMS) following the on-line combustion of compounds separated by GC has also become very popular as a method to guarantee the authenticity of food products based on the measurement of the isotope distribution at natural abundance level. Multidimensional gas chromatography (GC×GC) in combination with mass spectrometry is becoming also a very powerful technique in food analysis, and some applications dealing with frauds and authenticity aspects are also reported.

In this chapter, trends of GC-MS techniques to address food integrity and authenticity aspects for the identification and the prevention of frauds will be presented. Coverage of all kind of applications is beyond the scope of the present contribution. Therefore, the present chapter will focus on the most relevant applications of targeted and non-targeted GC-MS approaches published in the last years.

NON-TARGETED GC-MS FINGERPRINTING PROFILING APPROACHES

The use of chromatographic fingerprinting as an innovative approach to address food integrity and food authentication is increasing [10]. As previously commented in the introduction section, fingerprinting methods describe a variety of analytical methodologies that provide analytical signals related to the chemical composition of food products in a non-selective way (without assuming any previous knowledge of relevant or irrelevant food components). This information can be obtained by collecting data as a spectrum or a chromatogram by GC-MS.

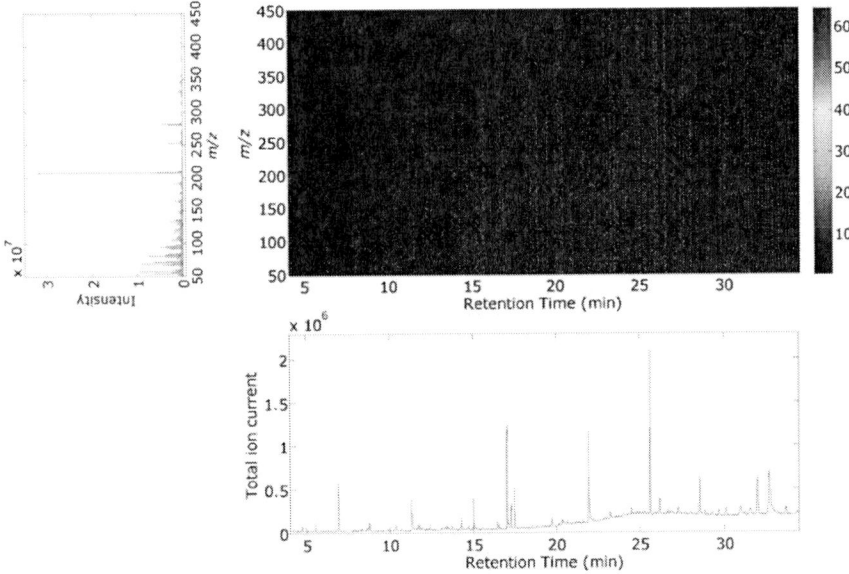

Figure 1. Image of the raw two-way GC-MS fingerprint profiling data set of a conventional grown basil sample. The integrated total ion chromatogram and the integrated mass spectrum are also provided. Reprinted with permission of reference [11]. Copyright (2013) American Chemical Society.

As an example, Wang et al. [11] proposed a method for the authentication of basils by GC-MS non-targeted chemical fingerprinting profiling. Basil plants cultivated by organic (ecological) and conventional farming practices were accurately classified by pattern recognition of GC-MS data. Samples were extracted by direct solvent extraction employing pentane/acetone (90:10 v/v) with the aim of obtaining a general fingerprint of major components. The GC separation was accomplished on a 30 m × 0.25 mm × 0.25 μm 5% diphenyl/95% dimethyl polysiloxane cross-linked capillary column under gradient temperature ramp. MS acquisition in a quadrupole analyzer was performed in full scan mode from m/z 50 to 450. Thus, the fingerprinting data sets were binned by retention time from 4.1 to 34.5 min with a 0.01 min increment and binned by mass-to-charge ratio from 50 to 450 Th with a 1 Th increment. Figure 1 shows the image of a raw two-way GC-MS data fingerprint profiling of a conventional basil sample, together with the total ion chromatogram and the integrated mass spectrum.

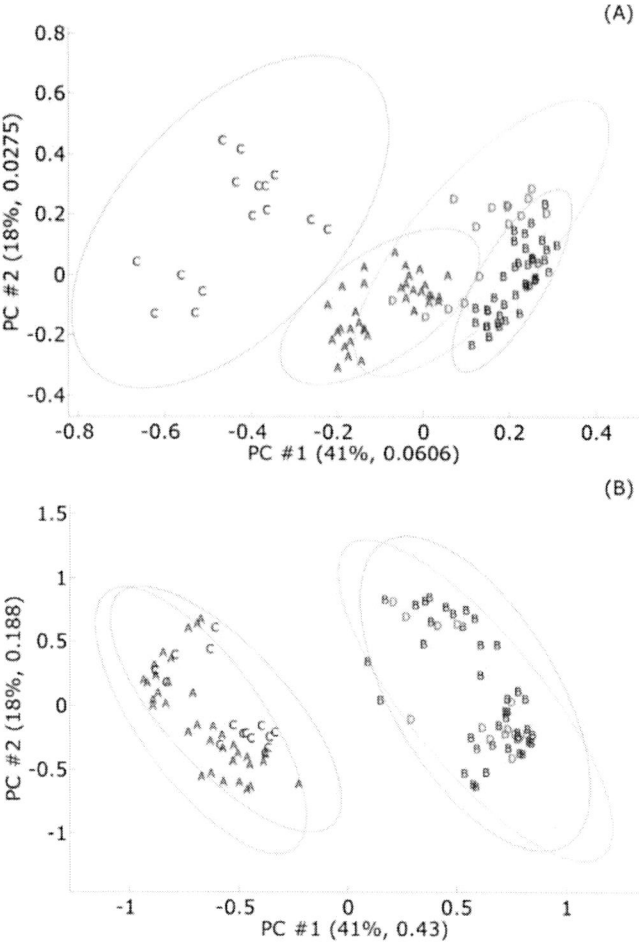

A: The training set of organic basils
B: The training set of conventional basils
C: The prediction set of organic basils
D: The prediction set of conventional basils

Figure 2. Principal component analysis score plots for two-way GC-MS data sets of conventional and organic basils before (A) and after (B) mean-centering the data sets. The percent variance spanned by the principal components is given in parentheses with the absolute variance. A 95% confidence interval is drawn around the mean of each class. The samples employed for the prediction set were collected 2.5 months after the ones employed in the training set. Reprinted with permission of reference [11]. Copyright (2013) American Chemical Society.

Prior to chemometric data processing, baseline correction and retention time alignment was performed. Variables used for the classification of samples (classifiers) were built with the two-way data sets obtained, the total ion chromatogram representation of the data sets, and the total mass spectrum representation of the data sets. Figure 2 shows the PCA score plots for the two-way GC-MS data sets of the conventional and organic basils analyzed before (A) and after (B) mean-centering the data sets.

The authors commented that after the initial experiments with the basil samples employed for the training set and collected 2.5 months before the ones for the prediction (validation) set, small deviations of the instrument status or manual operation caused inconsistencies between the new and the old data sets, as can be seen in Figure 2A, even although the employed experimental conditions were identical. However, with the correct data pretreatment results were improved. The model-building set and the prediction set were each individual mean-centered, and as can be seen in Figure 2B, the organic grown basils were all well separated from the conventional grown basils, and for each class the score of objects in the prediction set corresponded to the scores of the objects in the training set.

Volatolomics, the chemical analysis (detection and monitoring) of compounds associated with volatile metabolites in an organism is nowadays at the cutting edge of science, and is employed in a broad range of application fields including biomedical research, nutrition, toxicological analysis, forensics, safety and security [12]. The use of GC-MS techniques to study the volatile organic compounds (VOCs) that are "emitted" by food products is also becoming an useful strategy to address food integrity and authenticity, and several interesting applications can be found in the literature [13–20].

For instance, Vasta et al. [17] employed the determination of VOCs in raw meat for the authentication of the feeding background of farm animals. The authors employed dynamic headspace (DH) for the extraction of VOCs from raw lamb muscle, by performing the extraction at a temperature of 35 °C and using 6.25 g of sample. A total of 16 muscle samples from experimental lambs were analyzed, divided into concentrate-fed lambs (8 samples) and pasture-fed lambs (8 samples).

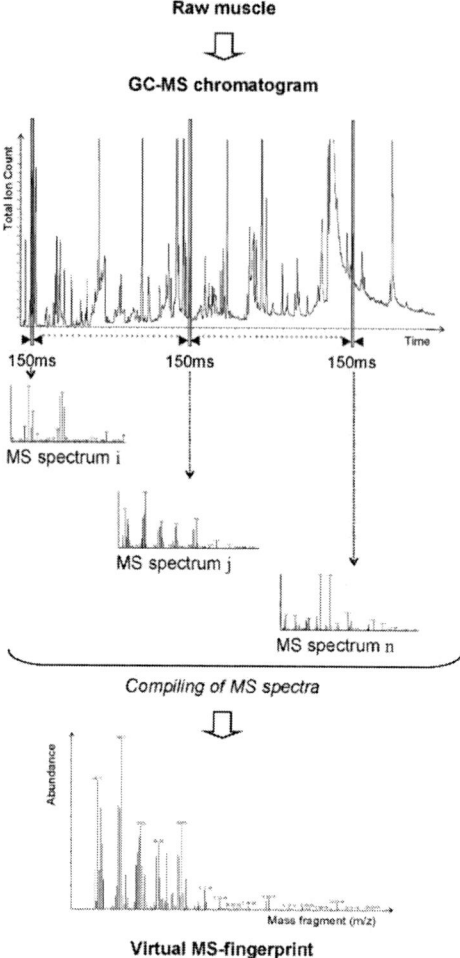

Figure 3. Scheme of the construction of a virtual MS fingerprint of the volatile fraction of a raw muscle sample from a DH-GC-MS chromatogram of this tissue. The mass spectra that were acquired every 150 ms of the GC-MS chromatogram were summed and then converted in a virtual MS spectral fingerprint characterized by the abundance of 198 mass fragments ranging from 33 to 230 amu. Reprinted with permission of reference [17]. Copyright (2007) American Chemical Society.

To carefully explore the information provided by the DH-GC-MS analysis, the MS spectra acquired along the chromatogram were summed and then converted in a virtual-DH-MS spectral fingerprint, as indicated in Figure 3, which was then employed as chemical descriptor to have a quick overview of the discriminative potential of the obtained volatile fraction.

The detailed examination of the information provided by the GC dimension showed that 33 VOCs among the 204 detected in the lamb muscle by DH-GC-MS enabled the authors to discriminate the type of feeding on the lambs by principal component analysis, as can be seen in Figure 4.

A similar approach was also employed by the same research group to determinate the animal feeding system (pasture or concentrate) by analyzing in parallel the volatile fraction of three adipose tissues excised from 16 lambs by DH-GC-MS [19]. Using a lipid liquid phase extraction, heating the ground tissue to 70 °C, was shown to be the best sample preparation mode before DH-GC-MS analysis to achieve a good representation of the starting material, while getting a good extraction and reproducibility. On the basis of growth rate and anatomical location, three different lamb adipose tissues were analyzed by the authors: perirenal fat (PRF), caudal subcutaneous fat (CSCF), and heart fat (HF).

The multivariate discriminant analyses results obtained by the authors confirmed that animal feeding discrimination was improved when PRF, CSCF, and HF adipose tissues were considered simultaneously, even if HF contributed with minimal information.

Marker discovery in volatolomics based on systematic alignment of DH-GC-MS signals was also applied to food authentication [16]. In this case, the authors demonstrated the relevance of a two-step alignment-based strategy for the systematic research of VOC markers. The first step consisted on reducing the time shifts with warping techniques like Correlation Optimized Warping (COW), followed by an accurate peak apex alignment in order to refine residual local misalignments and to enable further systematic research by chemometric data treatment.

The authors implemented this strategy on 117 GC-MS analysis of the volatolome of three walnut vegetable oils with very similar composition. 100% classification rate was observed by linear discriminant analysis (LDA) when the two-step alignment strategy was employed. The untargeted analysis enabled the authors to discover 184 VOCs as markers to discriminate the three groups of walnut oil samples.

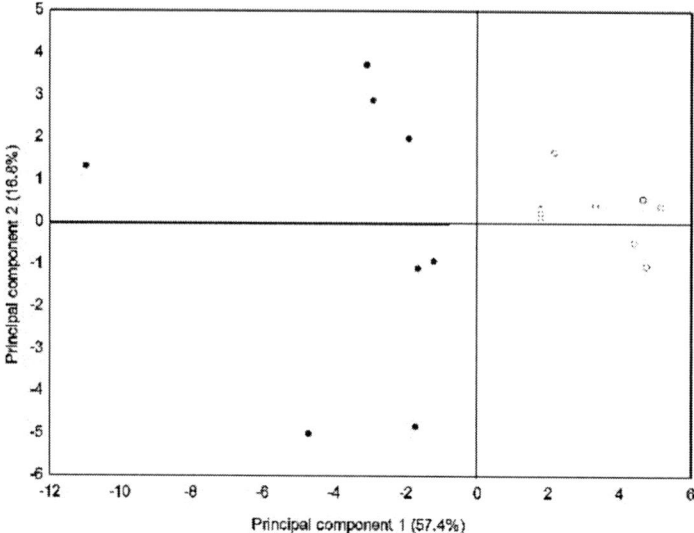

Figure 4. Discrimination of lamb muscle according to the type of feeding based on the GC-MS fingerprinting profile of their volatile fraction. Solid circles: concentrate-fed lambs; open circles: pasture-fed lambs. Reprinted with permission of reference [17]. Copyright (2007) American Chemical Society.

Very recent applications on this topic are the works of Mentana et al. whom employed volatolomics approaches to investigate the effects of mild preservation technology on perishable foods (ready-to-cook tuna-burgers) [14], and to discriminate olive tree varieties infected by *Sylella fastidiosa* [15]. In both cases, the volatolome was obtained by head-space solid-phase microextraction in combination with GC-MS (HS-SPME-GC-MS). Principal component analysis of the obtained HS-SPME-GC-MS volatolomic fingerprinting profiles allowed discriminating among two olive tree cultivars (*Olea europaea* L.) in Salento (Italy) area (*Ogliarola* and *Cellina di Nardó* cultivars), infected and non-infected by *Sylella fastidiosa* [15]. Ready-to-cook tuna-burgers were also perfectly discriminated by PCA of the obtained HS-SPME-GC-MS volatolome fingerprints among samples containing or not *Lactobacilus paracasei* (a protective microbial strain), and among samples packaged in a modified atmosphere packaging (5% O_2, 95% CO_2) or packaged in air [14].

Non-targeted screening (metabolomics fingerprinting) of (semi)volatile substances by gas chromatography-high resolution mass spectrometry (GC-

HRMS) using a quadrupole-time-of-flight (Q-TOF) analyzer was also described to assess the quality and authenticity of Scotch Whisky [21]. Two extraction and pre-concentration strategies were evaluated by the authors, SPME and ethyl acetate extraction, the latter one preferred because not only volatile, but also a number of semi-volatile compounds, were detected for sample characterization. 171 authentic whisky samples provided by the Scotch Whisky Research Institute and 20 fake whisky samples were analyzed. Fingerprinting data was then subjected to PCA and PLS-DA. The PLS-DA classification model constructed with the fingerprinting data obtained by analyzing the ethyl acetate extracts was able to discriminate whiskies according to the type of cask in which they were matured (bourbon versus bourbon and wine), and significant "markers" for the bourbon and wine cask maturation, such as N-(3-methylbutyl) acetamide and 5-oxooxolane-2-carboxylic acid, were identified. Discrimination among malt and blended whiskies was also accomplished, and several markers were also identified. Finally, the proposed method was also able to identify fake whiskies based on the differences obtained in the GC-HRMS fingerprints, and a number of synthetic flavorings such as ethyl vanillin were identified.

GAS CHROMATOGRAPHY COUPLED TO ISOTOPE RATIO MASS SPECTROMETRY

Compound specific isotope analysis by isotope ratio mass spectrometry (IRMS) following the on-line combustion (C) of compounds previously separated by GC is a relatively young analytical method. Gas chromatography coupled to isotope ratio mass spectrometry (GC-IRMS) has increasingly become one of the methods of choice in the authenticity control of foodstuffs due to its ability to measure isotope distribution at natural abundance level with great accuracy and precision [22]. The scope of this chapter is not to describe the principles of operation of GC-C-IRMS technique. This is very well addressed, together with the historical developments of the method, a discussion on the natural variability of carbon and nitrogen stable isotope ratios, and a deep overview of the applications of this technique for food and beverages authenticity and traceability, in a very recommended review by van Leeuwen et al. [23]. In this section, several application examples of GC-C-IRMS for the authentication of foodstuffs will be described.

For example, authenticity and traceability of vanilla flavors were investigated using GC-IRMS for the analysis of stable isotopes of carbon and hydrogen by Hansen et al. [24]. For that purpose, vanilla flavors produced by chemical synthesis, fermentation, and extracted from two different species of the vanilla orchid were extracted by maceration employing an ethanolic solution for 72 h, followed by a further extraction of 3 mL of the ethanol/water extract with 3 mL of ethyl acetate/cyclohexane (1:1). Separation of the sample extracts was performed on a Trace GC Ultra (Thermo Scientific) instrument fitted with a DB-5 capillary column (30 m × 0.250 mm inner diameter, with d_f of 0.25 μm). For the determination of $\delta^{13}C$, compounds eluting from the GC were combusted in an oxidation reactor (NiO tube with NiO/CuO/Pt) operated at 1000 °C, while for δ^2H determinations, a high-temperature pyrolysis reactor (HTC reactor), consisting of a ceramic tube with no catalyst and operated at 1420 °C was employed. As commented by the authors, Vanilla orchids primarily use the crassulacean acid metabolism for fixing CO_2, while the substrate mostly used for production of vanillin by fermentation is based on plants fixing CO_2 via the C3 metabolic pathway. The two ways of carbon fixation cause differences in the ratio of the heavy to light stable carbon isotope ($\delta^{13}C$) that is later incorporated into the vanillin molecule, and therefore, it can be used to differentiate natural vanillin from bio-vanillin. Synthetic vanillin made from petrochemicals are depleted in the heavy carbon isotopes and expected to show the lowest values of $\delta^{13}C$, while vanillin from the vanilla orchid would have the highest values. Vanillin produced by C3 plants is expected to have values of $\delta^{13}C$ in the same range as vanillin made from petrochemicals. In addition, the ratios of the heavy/light stable hydrogen isotope (δ^2H) in water precipitation differ over the world primarily because of longitudinal, altitudinal, and continental effects. As a result, the δ^2H value of water taken up by a plant depends upon the location of its habitat. Because the plant absorbs water and uses it for biosynthesis of metabolites, the values of δ^2H in the local precipitation will be reflected in these secondary metabolites, for example, vanillin from the vanilla orchid. Therefore, values of $\delta^{13}C$ and δ^2H can be used for studies of authenticity and traceability of vanilla flavor. Figure 5 shows a graphic representation of $\delta^{13}C$ versus δ^2H, revealing that vanillin extracted from pods grown in adjacent geographic origins grouped together.

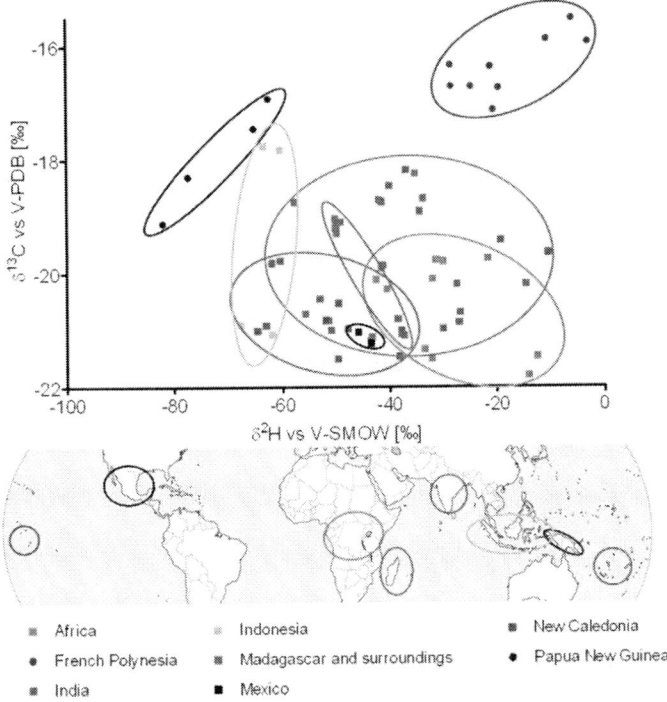

Figure 5. Results of δ^2H versus V-SMOW and $\delta^{13}C$ versus V-PDB for vanillin extracted from vanilla pods. Points are labeled with geographic origin. The geographical origins are separated by colors, with *V. tahitensis* represented by circles and *V. planifolia* represented by squares. The large circles in the scatter plot are color-correlated to the circles in the world map, showing the regions from where the samples originate. Reprinted with permission of reference [24]. Copyright (2014) American Chemical Society.

The combined use of the results of $\delta^{13}C$ and $\delta^{2}H$ analyses for vanillin, extracted from vanilla pods, shows a relatively close clustering pending upon species as well as geographic location of growth. Therefore, the analysis of δ values of carbon and hydrogen by compound-specific IRMS is a useful tool in the verification of the authenticity and traceability of vanilla pods.

Recently, C and H stable isotope ratio analysis using SPME and GC-IRMS was also employed for vanillin authentication [25]. The authors analyzed almost 50 authentic samples from vanilla pods, nature-identical (ex) and synthetic vanillin and 4 commercial food products containing vanillin. All the samples were subjected to HS-SPME-GC-IRMS analysis and most of them to GC-IRMS analysis after solvent extraction of vanillin. The SPME method developed by the authors for $^{2}H/^{1}H$ analysis guarantees the absence of isotopic fractionation, good repeatability and reproducibility and savings in terms of time (from 30 to 5 min) and solvent. HS-SPME GC-IRMS analysis of $\delta^{2}H$ and $\delta^{13}C$ can be proposed as a rapid and robust method to discriminate different types of vanillin and assess the authenticity of natural vanillin, also contained in food matrices.

GC-IRMS is also a widely employed technique in the authentication of alcoholic beverages [26–29], as well as alcoholic and non-alcoholic sparkling drinks [29]. For example, Aguilera-Cisneros et al. [27] proposed the use of HS-SPME-GC-IRMS, in the combustion (C) and the pyrolysis (P) modes (HS-SPME-GC-C/P-IRMS), to determine the $\delta^{13}C$ and $\delta^{18}O$ values of ethanol in authentic and commercial tequila samples, as well as a number of other spirits. In Mexico, the production of tequila is restricted by law to the blue agave (*A. tequilana* Weber var. Azul) as well as to specifically designated geographic areas, primarily the state of Jalisco in west-central Mexico [30].

As an example, Figure 6 shows (a) the correlation of $\delta^{13}C$ and $\delta^{18}O$ values of ethanol determined in authentic and commercial tequila samples and (b) the correlation of $\delta^{13}C$ and $\delta^{18}O$ values of ethanol determined in authentic and commercial tequila, as well as in other spirits. The authors concluded that despite the limited number of samples, the restriction to one production line, and the lack of control of fermentation conditions, the information provided especially by the $^{18}O/^{16}O$ ratio analysis of ethanol can be helpful to the industry and the control laboratories for the authenticity assessment of tequila.

Calderone et al. [29] proposed the direct analysis from the headspace of several sparkling and soft-drinks without any prior purification step to measure $^{13}C/^{12}C$ isotopic ratio of carbon dioxide by GC-IRMS as a straightforward technique to characterize these beverages. Authentication aspects regarding those sparkling beverages can then be easily addressed. For example, the origin of the sugar added in wine can be authenticate, because in fermented products carbon dioxide must reflect the botanical origin of sugars from which the gas is produced, according to the three different photosynthetic paths followed by plants in sugar synthesis: C3 or Calvin cycle [31], C4 or Hatch-Slack cycle [32] and Crassulacean Acid Metabolism (CAM) [33]. In the case of carbonated waters, $^{13}C/^{12}C$ isotopic ratio will help in identifying if it is a naturally carbonated water or, if on the contrary, CO_2 of an origin different than the water table or deposit from which the water comes was added.

Hattori et al. [34] developed a simple and rapid method able to discriminate among the fermentation of the raw materials employed in vinegar by measuring the hydrogen and carbon isotope ratio of acetic acid using head space solid-phase microextraction (HS-SPME) combined with gas chromatography-high temperature conversion or combustion-isotope ratio mass spectrometry (GC-TC/C-IRMS). A GC-conversion interface was used for hydrogen measurements, and a GC-combustion interface was used for carbon measurements.

Fourteen commercial vinegars were obtained by the authors from supermarkets in Japan, fermented from different raw materials such as rice, tomato, apple fruit juice, corn, or alcohol sake lees, among others. As an example, Figure 7 shows δD and $\delta^{13}C$ values of acetic acid in vinegars of different raw materials. As can be seen, the hydrogen and carbon isotope ratios were good parameters to discriminate the botanical origins of the acetic acid. In particular, the difference between C3 and C4 plants was clearly ascertained.

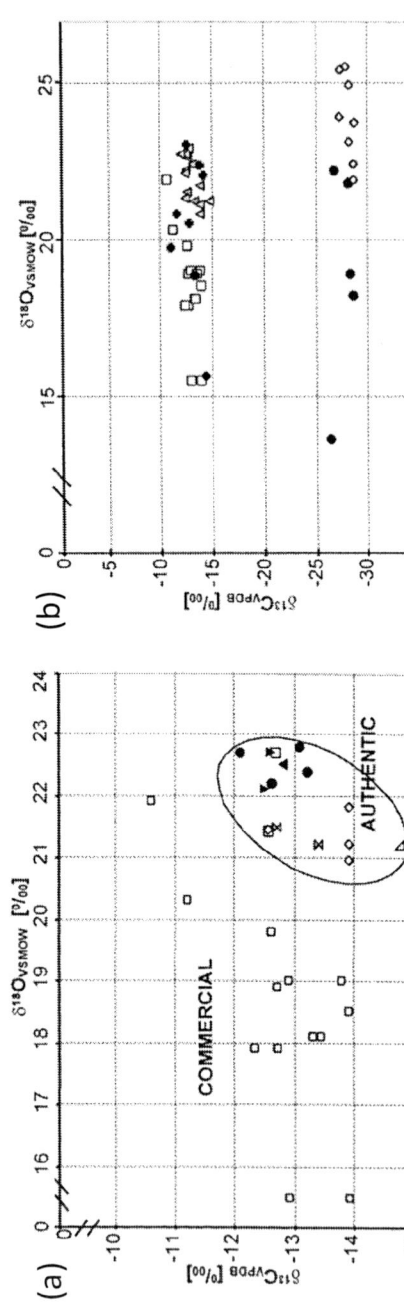

Figure 6. (a) Correlation of $\delta^{13}C$ and $\delta^{18}O$ data of ethanol determined by HS-SPME-GC-C/P-IRMS in authentic and commercial tequila samples (mean values from five determinations with standard deviations of 0.1–0.2 and 0.2–0.5 for $\delta^{13}C$ and $\delta^{18}O$, respectively). 100% agave: white, ■; rested, ●; aged, ▲. Mixed: white, (open right triangle); rested, ◊; aged, ✕; commercial tequila, □. (b) Correlation of $\delta^{13}C$ and $\delta^{18}O$ data of ethanol determined by HS-SPME-GC-C/P-IRMS in authentic and commercial tequilas as well as in other spirits (mean values from five determinations with standard deviations of 0.1–0.2 and 0.2–0.5 for $\delta13C$ and $\delta18O$, respectively). Commercial tequila, □; authentic tequila, (gray triangle); rum, (black cross); vodka, ●; wine distillates, ◊. Reprinted with permission of reference [27]. Copyright (2002) American Chemical Society.

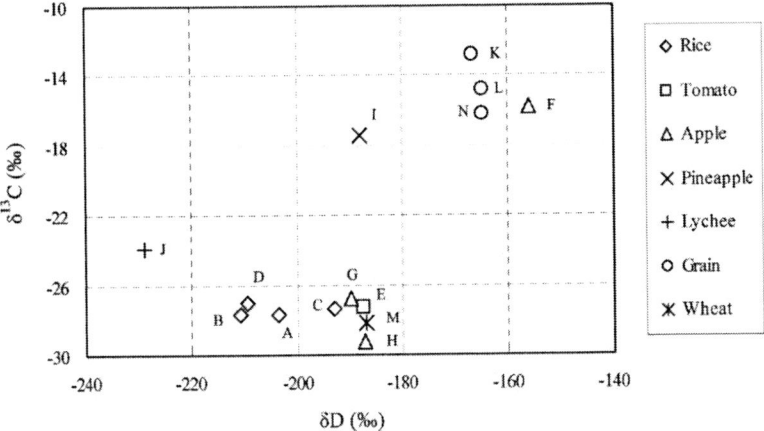

Figure 7. δD and δ^{13}C values of acetic acid in vinegars of different raw materials. Reprinted with permission of reference [34]. Copyright (2010) American Chemical Society.

The authencitiy assessment of flavoured strawberry foods was also accomplished by Schipilliti et al. [35] using HS-SPME-GC-C-IRMS. In this work, the authors investigated on the carbon isotope ratio of numerous selected aroma active volatile components (methyl butanoate, ethyl butanoate, hex-(2E)-enal, methyl hexanoate, buthyl butanoate, ethyl hexanoate, hexyl acetate, linalool, hexyl butanoate, octyl isovalerate, γ-decalactone and octyl hexanoate) of organic Italian fresh strawberries. They also compared the obtained results with those obtained by analyzing the HS-SPME extracts of commercial flavoured food matrices. The ^{13}C/^{12}C obtained values allowed the authors to differentiate between different biogenetic pathways (C3 and CAM plants), and more interestingly between plants of the same CO_2 fixation group (C3 plants).

In addition, the authors demonstrated that results obtained by the proposed GC–C-IRMS method were in agreement with the ones obtained with enantioselective gas chroamtography (Es-GC) by measuring the enantiomeric distribution of linalool and γ-decalactone, both techniques allowing to detect the presence of non-natural strawberry aromas in the food matrices studied.

TARGETED GC-MS APPROACHES

Another trend when employing GC-MS techniques to address food integrity and authenticity aspects is the use of targeted approaches. As previously commented in the introduction section, targeted approaches are based on the specific determination of a given group of known selected chemicals, or a group of chemicals belonging to the same family or with similar structural features. In this sense, profiling of lipids is frequently employed in the characterization and authentication of some food products [36], and GC-MS is among the techniques most frequently employed for that purpose.

For instance, Jurado et al. [37] proposed the characterization and quantitation of 4-methylsterols and 4,4-dimethylsterols from the Iberian pic subcutaneous fat by GC-MS and GC-flame ionization detector (GC-FID) as a way of authenticating the pig fattening system employed ("extensive" versus "intensive"). To perform the analysis, the sample lipids were extracted by melting the subcutaneous fat in a microwave oven. Then, the unsaponifiable matter was fractionated by thin layer chromatography. The GC separation was performed on a capillary SPB-5 column (30 m × 0.25 mm i.d., 0.15 mm film thickness). The full scan of free and trimethyl silyl ethers was used as acquisition mode. Six compounds were identified for the first time in this type of samples by the authors: (3b,4a,5a)-4-methyl-cholesta-7-en-3-ol, (3b,4a,5a)-4-methyl- cholesta-8(14)-en-3-ol, (3b,5a)-4,4-dimethyl-cholesta-8(14),24-dien-3-ol, (3b)-lanosta-8,24-dien-3-ol, (3b,5a)-4,4-dimethyl-cholesta-8,14-dien-3-ol and (3b)-lanost-9(11),24-dien-3-ol. These compounds were then employed as chemical descriptors to differentiate between the two pig-fattening systems by means of chemometric pattern recognition techniques, being able to achieve an overall classification performance of 91.7%.

Caligiani et al. [38] focused instead in the quantitative GC-MS determination of cyclopropane fatty acids (CPFA) in cheese, which were then employed as new molecular markers for the authentication of Parmigiano Reggiano cheese. CPFA, as lactobacillic acid and dihydrosterculic acid, are components of bacterial membranes and have been recently detected in milk and in dairy products from cows fed with corn silage. Therefore, its presence can be employed as marker of the use or not of silages and their presence in milk-processed products, such as cheese, can be employed to authenticate their production region according to specific production rules. The proposed method allowed to reach limits of detection and quantitation of CPFA of 60 and 200 mg/kg of cheese fat. The authors analyzed 304 samples of PDO cheeses of certified origin, including Parmigiano Reggiano (Italy), Grana Padano (Italy), Fontina (Italy), Comté (France), and Gruyère (Switzerland). Figure 8 shows the total GC-MS chromatogram of fatty acid methyl esters (FAMEs) of a Grana Padano cheese fat. An enlarged view of the CPFA peak elution zone, and the comparison between a positive and a negative CPFA sample are also included.

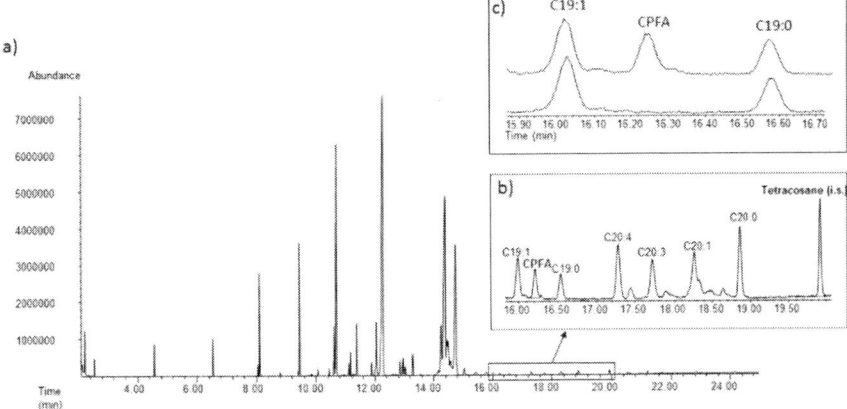

Figure 8. (a) Total chromatogram (gas chromatography–mass spectrometry) of FAMEs of a Grana Padano cheese fat, (b) enlarged view of CPFA peak elution zone, and (c) comparison between a sample positive and a sample negative to CPFA. Reprinted with permission of reference [38]. Copyright (2016) American Chemical Society.

The results obtained in these analysis revealed that CPFA were absent in all of the cheeses whose Production Specification Rules expressly forbid the use of silages (Parmigiano Reggiano, Fontina, Comté, and Gruyère). CPFA were instead present (300−830 mg/kg of fat) in all of the samples of Grana Padano cheese (where the use of silages was allowed). In consequence, CPFA can be proposed as good markers of silage feedings in cheese, as well as to address cheese production regions to guarantee PDO attributes.

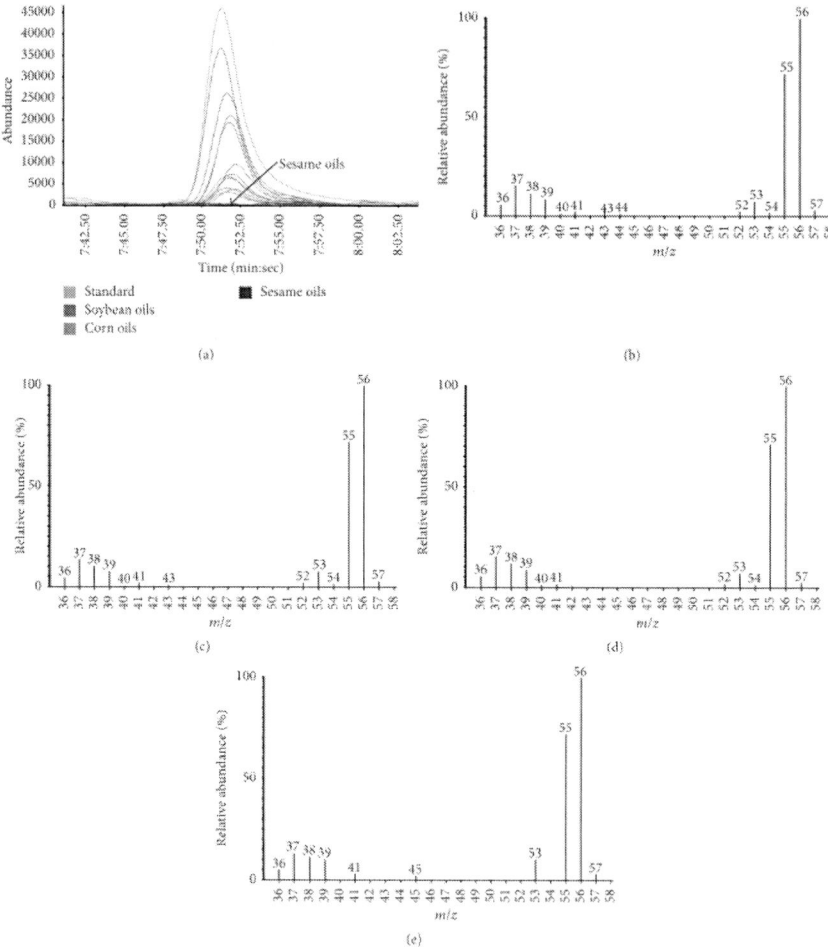

Figure 9. (a) Chromatograms of 2-propenal in six corn oils (blue line), five soybean oils (red line), and twelve sesame oils (black line) analyzed using the proposed GC-TOFMS method. The figure also include the MS spectra of 2-propenal in (b) corn oil, (c) soybean oil, (d) authentic standard, and (e) the Wiley library. All experimental MS spectra were registered at 70 eV of electron energy. Reproduced from the Open Access reference [39].

An example of a targeted GC-MS methodology for the authentication of foodstuffs by means of determining a specific sample bio-marker is the work described by Mansur et al. [39]. The authors proposed the determination of a simple marker, 2-propenal, by HS-SPME, under mild temperature conditions, coupled to GC-TOFMS for the authentication of unrefined sesame oil adulterated with refined corn or soybean oils. This is due to the fact that 2-propenal is one of the highly reactive aldehydes detected in thermally processed edible oils, formed from the constituents of glycerides such as the glycerol moieties of triacyl and diacyl glycerides, and it can also be generated by degradation of unsaturated fatty acid backbones [40]. As an example, Figure 9 shows the GC-TOFMS chromatogram of 2-propenal in several analyzed oil samples, as well as its MS spectrum in several samples. With the proposed methodology, 2-propenal was detected in all the tested refined corn and soybean oils but not in any of the analyzed unrefined sesame oil samples, showing the importance of this compound as a bio(marker) to guarantee the authentication of unrefined sesame oil samples.

Two-Dimensional GC Coupled with Mass Spectrometry

Multidimensional gas chromatography offers excellent separation efficiency for the advanced characterization of volatiles and semi-volatiles in food samples. Nowadays, is becoming a popular technique for non-target and target compound identification, and for fingerprinting profiling applied to food analysis [41, 42]. The main benefit of this technique to address especially compound identification is the use of different orthogonal separations and different hyphenation possibilities, such as the coupling with mass spectrometry, expanding the range of compounds that can be detected. Although the number of applications focusing on the use of these techniques in food authentication is still somewhat limited, some very interesting works can be found in the literature.

For example, multidimensional gas chromatography coupled to combustion-isotope ratio mass spectrometry/quadrupole MS with a low-bleed ionic liquid secondary column was recently proposed by Sciarrone et al. for the authentication of truffles and products containing truffle [43]. Truffles are among the most expensive foods available in the market, usually used as flavoring additives for their distinctive aroma. The most valuable species is *Tuber magnatum* Pico, better known as "Alba white truffle", in which bis(methylthio)methane is the key aroma compound. Due to the price and high quality of this product, it is important to develop robust analytical methodologies able to authenticate among products containing genuine (natural) white truffles or synthetic truffle aroma. With this aim, the authors proposed the measurement of $\delta^{13}C$ ratios by high-efficiency HS-SPME-MDGC-C-IRMS with simultaneous quadrupole MS detection for the evaluation of the bis(methylthio)methane truffle marker. The proposed multidimensional GC methodology allowed resolving the coelution problem of these compounds with other sample components, and in order to minimize the effect of column bleeding on $\delta^{13}C$ measurement, a medium polarity ionic liquid-based stationary phase was employed, instead of a the typical polyethylene glycol one, as the secondary column. The authors analyzed 24 genuine white truffles harvested in Italy. Figure 10 represents the $\delta^{13}C$ value ranges obtained based on the geographical origin of the analyzed genuine (natural) white truffles.

In addition, two commercial intact truffles and 14 commercial samples of pasta, sauce, olive oil, cream, honey, and fresh cheese flavored with truffle aroma were also analyzed by the authors, and the results from $\delta^{13}C$ measurement were evaluated in comparison with those of genuine "white truffle" range and commercial synthetic bis(methylthio)methane standard. Among the samples analyzed, the authors found that some of them presented $\delta^{13}C$ compatible with the natural white truffle range measured. In other cases, the use of low quality (probably not fresh) white truffle was detected, being the sample fortified by a synthetic flavor.

In another work, SPE-GC×GC-TOFMS was employed for the characterization and authentication of edible oils by free phytosterol profiling [44]. The free phytosterol profiles of peanut, soybean, rapeseed, and sunflower seed oils were established by the proposed methodology and employed as chemical descriptors to classify the four edible oils with the help of unsupervised (principal component analysis and hierarchical clustering analysis) and supervised (random forests) multivariate statistical methods. The two columns employed were a 30 m DB-5ms (0.25 mm I.D. × 0.25 µm film thickness, phenyl arylene polymer, Agilent Technologies), and a Rxi-17Sil MS with dimensions of 2 m × 0.15 mm I.D. × 0.15 µm film thickness, similar to 50% phenyl/50% dimethylpolysiloxane (silarylene) (Restek, USA). The obtained results showed that free phytosterol profiles of the analyzed edible oils perfectly discriminated the samples in four groups completely separated, and therefore, these profiles can be used as important markers of the oils studied. In addition, the authors simulated the adulteration of a peanut oil with soybean oil, and the obtained results also demonstrated that free phytosterol profiles can also be used to detect this adulteration down to 5% soybean oil adulterant levels.

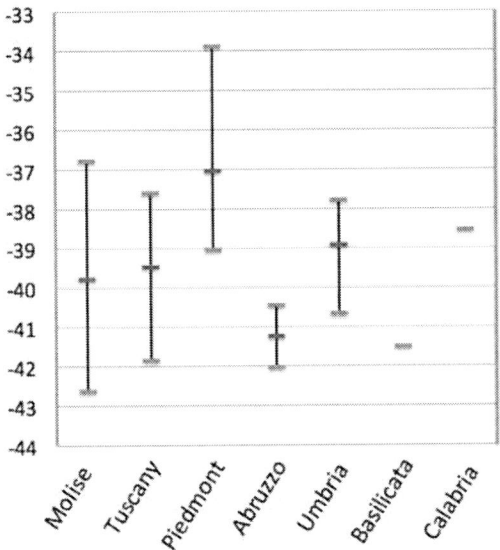

Figure 10. $\delta^{13}C$ ranges on the basis of the geographical origin of the genuine white truffle samples analyzed. Reprinted with permission of reference [43]. Copyright (2018) American Chemical Society.

SUMMARY AND CONCLUSION

The role of GC-MS methodologies to address food integrity and authenticity has been presented and discussed by means of some selected applications published in the last years. As in the case of other analytical methodologies employed for food analysis, two main approaches are typically used when GC-MS is applied to food authentication aspects: targeted and non-targeted strategies.

Among non-targeted GC-MS strategies, food volatolomics (the chemical analysis of compounds associated with the volatile metabolites of food products), is one of the methods of choice because GC is the best technique for the analysis of volatile compounds and volatile metabolites can be employed as good chemical descriptors of the origin of many food products. Therefore, food volatolome fingerprinting profiling obtained by GC-MS techniques is very useful to address the classification, characterization and authentication of natural and processed foodstuffs.

Isotope ratio mass spectrometry (IRMS) is also a very popular technique to address food authentication issues, especially those related with the origin of the food product. This is due to the fact that plants incorporate CO_2 following different metabolic pathways. Therefore, $\delta^{13}C/\delta^{12}C$ ratios, among other isotopic distributions, obtained by GC-IRMS are widely employed as markers to guarantee food authentication according to the production region, or the presence of natural versus synthetic products, etc., and several examples were described.

An interesting advantage of the non-targeted approaches (metabolomics fingerprinting) is that the chemical information obtained from the analyzed food samples is huge, becoming a very important source of information for the identification of biomarkers of the analyzed samples. Then, those identified biomarkers can be employed in targeted GC-MS strategies for the authentication of food products.

Finally, multidimensional GC-MS techniques have shown to be a very powerful technique in food analysis, and although the number of publications dealing with food integrity and authentication issues is still limited, sure it will increase in the near future.

ACKNOWLEDGMENTS

This work was supported by the Agency for Administration of University and Research Grants (Generalitat de Catalunya, Spain) under the projects 2017 SGR-310.

REFERENCES

[1] W. Van Rijswijk, L. J. Frewer, Consumer needs and requirements for food and ingredient traceability information, *Int. J. Consum. Stud.* 36 (2012) 282–290. doi:10.1111/j.1470-6431.2011.01001.x.

[2] COUNCIL REGULATION (EEC) No 2081/92 of 14 July 1992 on the protection of geographical indications and designations of origin for agricultural products and foodstuffs, Off. *J. European Communities* L208 1-8.

[3] J. C. Moore, J. Spink, M. Lipp, Development and Application of a Database of Food Ingredient Fraud and Economically Motivated Adulteration from 1980 to 2010, *J. Food Sci.* 77 (2012) R118–R126. doi:10.1111/j.1750-3841.2012.02657.x.

[4] C. M. E. Gossner, J. Schlundt, P. Ben Embarek, S. Hird, D. Lo-Fo-Wong, J. J. O. Beltran, K.N. Teoh, A. Tritscher, The melamine incident: Implications for international food and feed safety, *Environ. Health Perspect.* 117 (2009) 1803–1808. doi:10.1289/ehp.0900949.

[5] C. Lazzerini, C. Lazzerini, M. Cifelli, V. Domenici, Pigments in Extra-Virgin Olive Oils: Authenticity and Quality, Chapter 6 in D. Boskou and M. Clodoveo (Eds) *"Products from Olive Tree"*, Intech-Open, London, UK, (2016), pp. 99-114. doi:http://dx.doi.org/10.5772/57353.

[6] R. C. Alves, M. F. Barroso, M. B. González-García, M. B. P. P. Oliveira, C. Delerue-Matos, New Trends in Food Allergens Detection: Toward Biosensing Strategies, *Crit. Rev. Food Sci. Nutr.* 56 (2016) 2304–2319. doi:10.1080/10408398.2013.831026.

[7] G. P. Danezis, A. S. Tsagkaris, V. Brusic, C. A. Georgiou, Food authentication: state of the art and prospects, *Curr. Opin. Food Sci.* 10 (2016) 22–31. doi:10.1016/j.cofs.2016.07.003.

[8] S. Medina, R. Perestrelo, P. Silva, J. A. M. Pereira, J. S. Câmara, Current trends and recent advances on food authenticity technologies and chemometric approaches, *Trends Food Sci. Technol.* 85 (2019) 163–176. doi:10.1016/j.tifs.2019.01.017.

[9] S. D. Brown, R. Tauler, B. Walczak. *Comprehensive Chemometrics. Chemical and Biochemical Data Analysis*, Volume 3, Elsevier 2009, Amsterdam, The Netherlands.

[10] L. Cuadros-Rodríguez, C. Ruiz-Samblás, L. Valverde-Som, E. Pérez-Castaño, A. González-Casado, Chromatographic fingerprinting: An innovative approach for food "identitation" and food authentication - A tutorial, *Anal. Chim. Acta.* 909 (2016) 9–23. doi:10.1016/j.aca.2015.12.042.

[11] Z. Wang, P. Chen, L. Yu, P. B. De Harrington, Authentication of organically and conventionally grown basils by gas chromatography/mass spectrometry chemical profiles, *Anal. Chem.* 85 (2013) 2945–2953. doi:10.1021/ac303445v.

[12] S. Giannoukos, A. Agapiou, B. Brkić, S. Taylor, Volatolomics: A broad area of experimentation, *J. Chromatogr. B Anal. Technol. Biomed. Life Sci.* 1105 (2019) 136–147. doi:10.1016/j.jchromb.2018.12.015.

[13] L. Pillonel, S. Ampuero, R. Tabacchi, J.O. Bosset, Analytical methods for the determination of the geographic origin of Emmental cheese: Volatile compounds by GC/MS-FID and electronic nose, *Eur. Food Res. Technol.* 216 (2003) 179–183. doi:10.1007/s00217-002-0629-4.

[14] A. Mentana, A. Conte, M. A. Del Nobile, M. Quinto, D. Centonze, Investigating the effects of mild preservation technology on perishable foods by volatolomics: The case study of ready-to-cook tuna-burgers, Lwt. 115 (2019) 108425. doi:10.1016/j.lwt.2019.108425.

[15] A. Mentana, I. Camele, S. M. Mang, G. E. De Benedetto, S. Frisullo, D. Centonze, Volatolomics approach by HS-SPME-GC-MS and multivariate analysis to discriminate olive tree varieties infected by

Xylella fastidiosa, *Phytochem. Anal.* (2019) 623–634. doi:10.1002/pca.2835.

[16] S. Abou-el-karam, J. Ratel, N. Kondjoyan, C. Truan, E. Engel, Marker discovery in volatolomics based on systematic alignment of GC-MS signals: Application to food authentication, *Anal. Chim. Acta.* 991 (2017) 58–67. doi:10.1016/j.aca.2017.08.019.

[17] V. Vasta, J. Ratel, E. Engel, Mass spectrometry analysis of volatile compounds in raw meat for the authentication of the feeding background of farm animals, *J. Agric. Food Chem.* 55 (2007) 4630–4639. doi:10.1021/jf063432n.

[18] D. Mannaş, T. Altuğ, SPME/GC/MS and sensory flavour profile analysis for estimation of authenticity of thyme honey, *Int. J. Food Sci. Technol.* 42 (2007) 133–138. doi:10.1111/j.1365-2621.2006.01157.x.

[19] G. Sivadier, J. Ratel, F. Bouvier, E. Engel, Authentication of meat products: Determination of animal feeding by parallel GC-MS analysis of three adipose tissues, *J. Agric. Food Chem.* 56 (2008) 9803–9812. doi:10.1021/jf801276b.

[20] B. S. Radovic, M. Careri, A. Mangia, M. Musci, M. Gerboles, E. Anklam, Contribution of dynamic headspace GC-MS analysis of aroma compounds to authenticity testing of honey, *Food Chem.* 72 (2001) 511–520. doi:10.1016/S0308-8146(00)00263-6.

[21] M. Stupak, I. Goodall, M. Tomaniova, J. Pulkrabova, J. Hajslova, A novel approach to assess the quality and authenticity of Scotch Whisky based on gas chromatography coupled to high resolution mass spectrometry, *Anal. Chim. Acta.* 1042 (2018) 60–70. doi:10.1016/j.aca.2018.09.017.

[22] W. Meier-Augenstein, Applied gas chromatography coupled to isotope ratio mass spectrometry, *J. Chromatogr. A.* 842 (1999) 351–371. doi:10.1016/S0021-9673(98)01057-7.

[23] K. A. van Leeuwen, P. D. Prenzler, D. Ryan, F. Camin, Gas Chromatography-Combustion-Isotope Ratio Mass Spectrometry for Traceability and Authenticity in Foods and Beverages, *Compr. Rev. Food Sci. Food Saf.* 13 (2014) 814–837. doi:10.1111/1541-4337.12096.

[24] A. M. S. Hansen, A. Fromberg, H. L. Frandsen, Authenticity and traceability of vanilla flavors by analysis of stable isotopes of carbon and hydrogen, *J. Agric. Food Chem.* 62 (2014) 10326–10331. doi:10.1021/jf503055k.

[25] M. Perini, S. Pianezze, L. Strojnik, F. Camin, C and H stable isotope ratio analysis using solid-phase microextraction and gas chromatography-isotope ratio mass spectrometry for vanillin authentication, *J. Chromatogr. A.* 1595 (2019) 168–173. doi:10.1016/j.chroma.2019.02.032.

[26] C. Bauer-Christoph, H. Wachter, N. Christoph, A. Roßmann, L. Adam, Assignment of raw material and authentication of spirits by gas chromatography, hydrogen- and carbon-isotope ratio measurements. I. Analytical methods and results of a study of commercial products, *Eur. Food Res. Technol.* 204 (1997) 445–452. doi:10.1007/s00217 0050111.

[27] B. O. Aguilar-Cisneros, M. G. López, E. Richling, F. Heckel, P. Schreier, Tequila authenticity assessment by headspace SPME-HRGC-IRMS analysis of 13C/12C and 18O/16O ratios of ethanol, *J. Agric. Food Chem.* 50 (2002) 7520–7523. doi:10.1021/jf0207777.

[28] C. Bauer-Christoph, N. Christoph, B. O. Aguilar-Cisneros, M. G. López, E. Richling, A. Rossmann, P. Schreier, Authentication of tequila by gas chromatography and stable isotope ratio analyses, *Eur. Food Res. Technol.* 217 (2003) 438–443. doi:10.1007/s00217-003-0782-4.

[29] G. Calderone, C. Guillou, F. Reniero, N. Naulet, Helping to authenticate sparkling drinks with 13C/12C of CO2 by gas chromatography-isotope ratio mass spectrometry, *Food Res. Int.* 40 (2007) 324–331. doi:10.1016/j.foodres.2006.10.001.

[30] Secofi, Norma oficial mexicana, NOM-006-SCFI-1993. Bebidas alcoho ́licas, tequila. In: *Diario oficial*; Mexico, 1993; p 48.

[31] A. M. Calvin, J. A. Bassham, *The photosynthesis of carbon compounds.* New York: Benjamin, 1962.

[32] M. D. Hatch, C. R. Slack, NoPhotosynthetic CO2-fixation pathways, *Annu. Rev. Plant Physiol.* 21 (1970) 141–162.

[33] T. Whelan, W. M. Sackett, C. R. Benedict, Enzymatic fractionation of carbon isotopes by phosphoenolpyruvate carboxylase from C4 plants, *Plant Physiol.* 51 (1973) 1051.

[34] R. Hattori, K. Yamada, H. Shibata, S. Hirano, O. Tajima, N. Yoshida, Measurement of the isotope ratio of acetic acid in vinegar by HS-SPME-GC-TC/C-IRMS, *J. Agric. Food Chem.* 58 (2010) 7115–7118. doi:10.1021/jf100406y.

[35] L. Schipilliti, P. Dugo, I. Bonaccorsi, L. Mondello, Headspace-solid phase microextraction coupled to gas chromatography-combustion-isotope ratio mass spectrometer and to enantioselective gas chromatography for strawberry flavoured food quality control, *J. Chromatogr. A.* 1218 (2011) 7481–7486. doi:10.1016/j.chroma. 2011.07.072.

[36] J. M. Bosque-Sendra, L. Cuadros-Rodríguez, C. Ruiz-Samblás, A. P. de la Mata, Combining chromatography and chemometrics for the characterization and authentication of fats and oils from triacylglycerol compositional data-A review, *Anal. Chim. Acta.* 724 (2012) 1–11. doi:10.1016/j.aca.2012.02.041.

[37] J. M. Jurado, A. Jiménez-Lirola, M. Narváez-Rivas, E. Gallardo, F. Pablos, M. León-Camacho, Characterization and quantification of 4-methylsterols and 4,4-dimethylsterols from Iberian pig subcutaneous fat by gas chromatography-mass spectrometry and gas chromatography-flame ionization detector and their use to authenticate the fattening systems, *Talanta.* 106 (2013) 14–19. doi:10.1016/ j.talanta.2012.12.006.

[38] A. Caligiani, M. Nocetti, V. Lolli, A. Marseglia, G. Palla, Development of a Quantitative GC-MS Method for the Detection of Cyclopropane Fatty Acids in Cheese as New Molecular Markers for Parmigiano Reggiano Authentication, *J. Agric. Food Chem.* 64 (2016) 4158–4164. doi:10.1021/acs.jafc.6b00913.

[39] A. R. Mansur, T. G. Nam, H. W. Jang, Y. S. Cho, M. Yoo, D. Seo, J. Ha, Determination of 2-Propenal Using Headspace Solid-Phase Microextraction Coupled to Gas Chromatography-Time-of-Flight

Mass Spectrometry as a Marker for Authentication of Unrefined Sesame Oil, *J. Chem.* 2017 (2017). doi:10.1155/2017/9106409.

[40] J. F. Stevens, C. S. Maier, Acrolein: Sources, metabolism, and biomolecular interactions relevant to human health and disease, Mol. Nutr. Food Res. 52 (2008) 7–25. doi:10.1002/mnfr.200700412.

[41] C. Cordero, J. Kiefl, P. Schieberle, S. E. Reichenbach, C. Bicchi, Comprehensive two-dimensional gas chromatography and food sensory properties: Potential and challenges, *Anal. Bioanal. Chem.* 407 (2015) 169–191. doi:10.1007/s00216-014-8248-z.

[42] Y. Nolvachai, C. Kulsing, P. J. Marriott, Multidimensional gas chromatography in food analysis, *TrAC - Trends Anal. Chem.* 96 (2017) 124–137. doi:10.1016/j.trac.2017.05.001.

[43] D. Sciarrone, A. Schepis, M. Zoccali, P. Donato, F. Vita, D. Creti, A. Alpi, L. Mondello, Multidimensional Gas Chromatography Coupled to Combustion-Isotope Ratio Mass Spectrometry/Quadrupole MS with a Low-Bleed Ionic Liquid Secondary Column for the Authentication of Truffles and Products Containing Truffle, *Anal. Chem.* 90 (2018) 6610–6617. doi:10.1021/acs.analchem.8b00386.

[44] B. Xu, L. Zhang, H. Wang, D. Luo, P. Li, Characterization and authentication of four important edible oils using free phytosterol profiles established by GC-GC-TOF/MS, *Anal. Methods.* 6 (2014) 6860–6870. doi:10.1039/c4ay01194e.

In: Gas Chromatography
Editor: Percy Henrichon

ISBN: 978-1-53617-350-5
© 2020 Nova Science Publishers, Inc.

Chapter 4

GAS CHROMATOGRAPHY–MASS SPECTROMETRY ANALYSIS OF SUGARCANE VINASSE

Mohamed A. Fagier[1,*] *and Mona O. Abdalrhman*[2]

[1]Department of Chemistry, Education Faculty,
University of Blue Nile, Ad-Damazin, Sudan
[2]Department of Chemistry, Faculty of Science and Arts,
Al Baha University, Al Bahah, Saudi Arabia

ABSTRACT

Structural analysis of wastewater is one of the important issue in wastewater treatment plant to select suitable and effective methods of treatments. Vinasse is a byproduct of ethanol and poses long-term risk to public health because of their persistent and toxic nature. Vinasse as distillery wastewater is considered as complex matrices, therefore, this study, aimed to find the organic compounds of vinasse by using gas chromatography- mass spectrometry GC-MS.

---

[*] Corresponding Author's E-mail: mfagery@yahoo.com

The solvent extraction method was used for sample preparation. Hexane and dichloromethane (DCM) were used as solvent. About fifteen organic compounds were detected and confirmed, twelve of them were identified and confirmed by hexane extraction, while only three compounds were identified and confirmed by DCM extraction. In general, the most abundant phenolic compounds were: 4-ethyl-3-methoxy phenol, 2,6- dimethoxy phenol, 4-allyl-2,6-dimethoxy phenol and (3,4-dimethoxyphenoxy) trimethylsilane.

Keywords: Sugarcane vinasse, phenolic compounds, carboxylic acids, GC-MS

1. INTRODUCTION

The World is progressively developing especially in industries sectors. Releasing of industries wastewater into the environment causes excessive hazards to the environment. Among the industrial wastewater, vinasse a byproducts of ethanol has a high content of organic composites with acidity nature. Vinasse is considered as the most hazardous industrial wastewater, due to the dark brown color, high biochemical oxygen demand (BOD) and chemical oxygen demand (COD) (Satyawali and Balakrishnan 2008; Fagier et al. 2016). Vinasse as one of the main sources of soil, rivers, surface and ground water pollution (Mohan et al. 2009).The recalcitrant nature and various characteristics of vinasse (Espana-Gamboa 2011) make the control of vinasse pollution is very challenging. Moreover the colored compound in vinasse has antioxidant properties and becomes toxic to all living systems including microorganisms (Soni Tiwari 2012).

In contraste, vinasse has many of utilization due to its high load of organic matter such as being in fertirrigation practices, it may represent a key reason in enhancing profitability and environmental outcomes of a sugarcane-to-ethanol plant (Cristiano E. Rodrigues Reis and Bo Hu 2017). Molina-Cortés et al., 2019 reports that vinasse has antioxidant properties, which would represent a new alternative for the usage and valorization of this material, previously considered as industrial waste. Therefore, identification of organic compounds of vinasse needs more attention to attenuate environmental concerns as well as utilizing and valorization of vinasse.

Substantial efforts have been explaining to find the composition of various types of vinasse (Dowd et al. 1994). Many researchers reported that the most important organic components of sugar cane vinasse are ethanol, acetic acid, glycerol, lignin and lactic acid (Dowd et al. 1994; Decloux and Bories 2002; Fagier et al. 2016), In addition to these low molecular weight compounds, vinasses may contain melanoidins, humic acid and phenolic compounds (Fitz Gibbon 1998). On the other hand, recently many scientists and researchers have compensated attention to study vinasse organic matter quality and mineralization potential using ionic chromatography (Parnaudeau 2008), also Benke (1997) and Emmanuel (2009) studied the characterizations of organic matter of vinasse using spectroscopic techniques.

Hyphenated techniques such as gas chromatography- mass spectrometry GC-MS can give a detailed chromatography profile of the sample and consequent measurements of the relative or absolute amounts of the components. The number of components measured will depend on the explanation of the chromatographic system and the explanation of the detection technique (John et al. 2004), therefore GC-MS is suitable technique for vinasse analysis

The aim of this work is to find the chemical ingredients mainly organic matter of sugarcane vinasse.in order to choose proper methods of treatment and/or a possible use of vinasse.

2. MATERIAL AND METHODS

2.1. Chemicals

Hexane, Sodium sulfate and sulfuric acid (98%) purchased from (Merck, Germany), Dichloromethane (DCM) obtained from (Suprasolv, Germany). N, O-bis trimethylsiyl trifluoroacetamide (BSTFA), Trimethylchlorosilane (TMCS) obtained from (supelco, USA). Sodium hydroxide purchased from (Fluka, Germany). Ultrapure water (Elga, USA) used for explaining of all aqueous solutions.

2.2. Collection of Samples

The vinasse used in this work obtained from Kenana sugar company, an ethanol distillery at White Nile State, Sudan. Sample collected directly after the distillation. The main chemical characteristics of the vinasse determined according to the Standard Methods for Examination of Water and Wastewater (Clesceri 1998).

2.3. Sample Preparation

The sample of vinasse was prepared for GC-MS analysis by solvent extraction method. Hexane and DCM used as a solvents. The sample and the solvent (1:4) shaken, manually, for 5 mints. The solvent removed from organic layer by a rotary evaporator (37°C). Extracted organic layer reduced under a gentle nitrogen stream (about 50min). TMCS added to DCM extraction, then heated to 70°C for 4 hr. and dried under nitrogen stream, then stored in fridge until GCMS analysis.

2.4. GC-MS Analysis

The analyses of the extract performed by gas chromatography mass spectrometry. These analyses carried out on a Hewlett- Packard Model 6890 gas chromatograph with splitless injector and a VB-5 5% phenyl-methylepolysiloxane column (30m Length, 0.25mm I.D., 0.25μm film thickness) equipped with a Hewlett- Packard Model 6890 mass selective detector give a HP ChemStation data acquisition system. Helium (purity 99.999%) used as a carrier gas. The chromatographic conditions are present in table (1). The data for analysis acquired from electron impact (EI) mode

70 (eV), scanning from 50-550 amu at 1.5sec/scan. Figures (1) and (14) shows the total ion chromatogram of extracted compounds by hexane and DCM respectively, the peaks numbered as retention time (R.T). Tables (2) and (3) summarize the identified compounds which extracted by hexane and DCM respectively.

3. RESULTS AND DISCUSSION

The GC-MS analysis of various compounds from vinasse were extracted by hexane and DCM. The identification of unknown compounds was initially accomplished by comparison with the MS library (NIST) and comforted by using Chemo bio draw program version ultra 11.0, and/or by interpreting the fragmentation pattern of the mass spectra. The comparison of the mass spectrums with the data base on MS library gave about 80% - 95% match as well as confirmatory compound structure match.

Table 1. GC-MS conditions

Oven Temperature program	Initial oven temperature 60°C, hold for 2 minutes; then up to 280°C at 6°C/min, then held at 280°C for 20 min
Gas flow rates	1.2 ml/min
Injection port temperature	290°C
Injection mode	Splitless (1 min) (1.0-1.4 µl; hot needle technique)
Column inlet pressure	10.4 pis
Average Velocity	40cm/s
Temperature of transfer line	300°C
Solvent delay	4 min

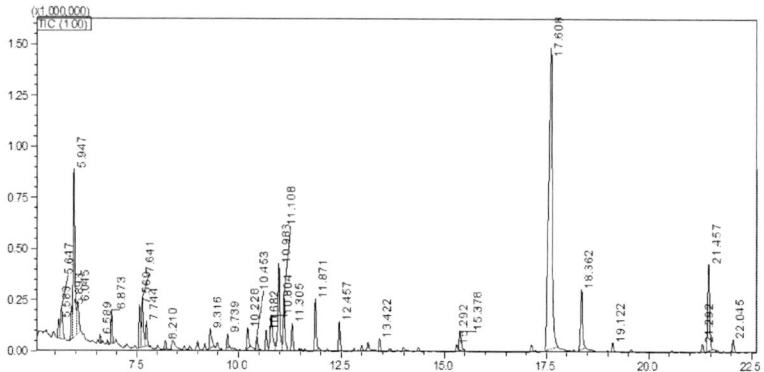

Figure 1. Total ion chromatogram of vinasse extracted by hexane. Peak numbers as R.T which is refer to table 2.

3.1. Identification of Hexane Extractable

The typical total ion chromatograms (TIC) of hexane extractable were given in figure (1), and table (2) represent the compounds extracted by hexane.

3.2. Identification of 2-Phenyl Ethanol

The EI mass spectrum of 2-phenyl ethanol. MW 122 is given in figure (2) the base peak is found at m/z 91 corresponding to M $[C_7H_7]^+$ result from the loss of $[CH_3O]$ the 2-phenyl ethanol appeared at R.T of 5.647 in total ion chromatogram.

3.3. Identification of 4-Ethyl-3-Methoxy Phenol

The EI mass spectrum of 4-ethyl-3-methoxy phenol MW 152 is given in figure (3). The base peak is found at m/z 137 corresponding to M $[C_8H_9O_2]^+$ due to the loss of $[CH_3]$. The 4-ethyl-3-methoxy phenol appeared at R.T 6.873 in total ion chromatogram.

Table 2. Identification and toxicity prediction of compounds extracted by hexane

R.T.	Name	Structure	M.W	Area %	Class
5.647	2-phenyl ethanol		122.16	2.23	Alcohol
6.873	4-ethyl- 3 methoxy phenol		152.19	1.42	Phenol
7.575	2,6-dimethoxy phenol		154.16	2.47	Phenol
9.742	1,2,3- triethoxy-5methylbenzene		182.22	0.64	Aromatic (benzene ring)

Table 2. (Continued)

R.T.	Name	Structure	M.W	Area %	Class
10.225	Dodecanoic acid		200.32	1.16	Carboxylic acid
10.675	3,4,5- trimethoxy phenol.		184.19	1.12	Phenol
10.983	4-allyl-2,6- dimethoxy phenol.		194.23	5.59	Phenol

R.T.	Name	Structure	M.W	Area %	Class
11.300	(E) -1-(3-hydroxy-2,6,6- trimethyl cyclohex-1- enyl)but-2-en-1-one.		208.30	1.43	Ketone
13.150	1,1,4,4-tetramethyl-2,5-dimethylenecyclohexane		164.29	0.70	Cycloalkane
17.608	Methyl palmitate		270.45	37.23	Ester
18.367	Palmitic acid.		256.42	4.12	Carboxylic acid
22.050	Methyl stearate		298.5	0.73	Ester

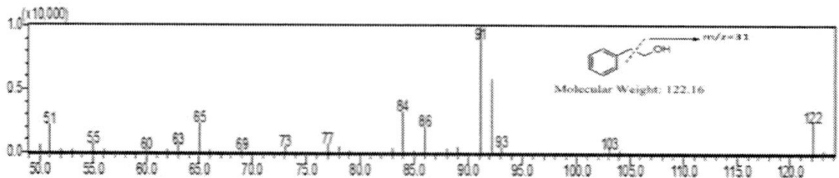

Figure 2. The mass spectrum analysis of 2-phenyl ethanol.

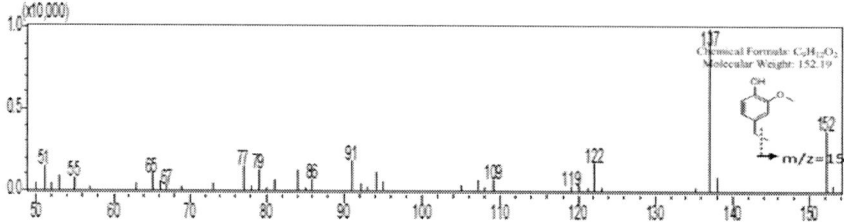

Figure 3. The mass spectrum analysis of 4-ethyl-3- methoxy phenol.

3.4. Identification of 2,6- Dimethoxy Phenol

The EI mass spectrum of 2,6- dimethoxy phenol MW 154 is given in figure (4) the base peak is found at m/z 139 corresponding to $M[C_7H_7O_3]^+$ due to the loss of [CH_3]. 2,6- dimethoxy phenol appeared at R.T 7.575 in total ion chromatogram.

3.5. Identification of 1,2,3- Triethoxy-5-Methyl Benzene

The EI mass spectrum of 1,2,3- triethoxy-5-methyl benzene MW 182 is given in figure (5). The MW of 183 probably due to the isotope of C13. The base peak is found at m/z 167 corresponding to $M[C_9H_{11}O_3]^+$ due to the loss of [CH_3]. 1,2,3- triethoxy-5-methyl benzene appeared at R.T 9.742 in total ion chromatogram.

3.6. Identification of Dodecanoic Acid

The EI mass spectrum of Dodecanoic acid MW 200 is given in figure (6) the loss of $[C_2H_5]$ results in $M[C_{10}H_{19}O_2]^+$ at m/z 171. While the base peak is found at m/z 73, Dodecanoic acid appeared at R.T 10.225 in total ion chromatogram.

3.7. Identification of 3,4,5- Trimethoxy Phenol

The EI mass spectrum of 3,4,5- trimethoxy phenol MW 184 is given in figure (7). The base peak is found at m/z 169 corresponding to $M[C_8H_9O_4]^+$. Due to the loss of $[CH_3]$. The 3,4,5- trimethoxy phenol appeared at R.T 10.675 in total ion chromatogram.

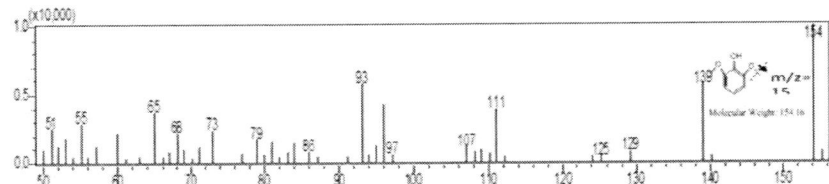

Figure 4. The mass spectrum analysis of 2,6- dimethoxy phenol.

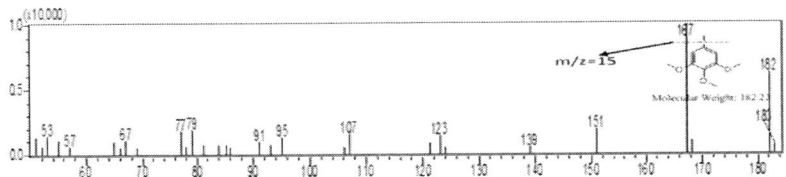

Figure 5. The mass spectrum analysis of 1,2,3- triethoxy-5-methyl benzene.

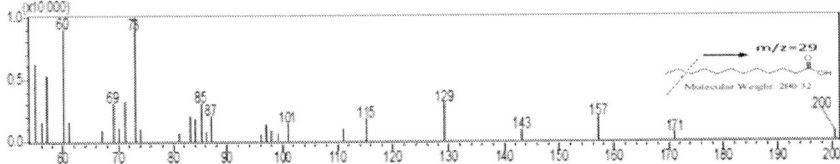

Figure 6. The mass spectrum analysis of Dodecanoic acid.

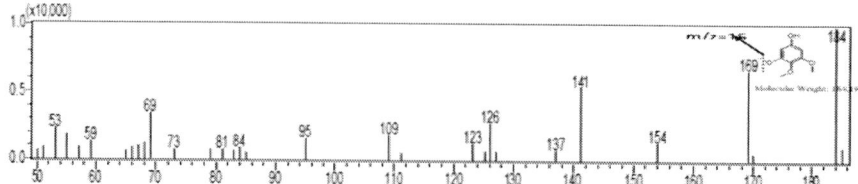

Figure 7. The mass spectrum analysis of 3,4,5- trimethoxy phenol.

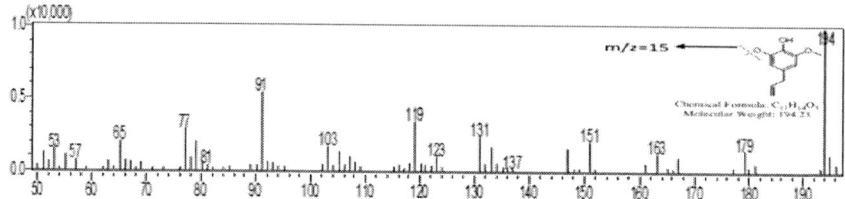

Figure 8. The mass spectrum analysis of 4-allyl-2,6- dimethoxy phenol.

3.8. Identification of 4-Allyl-2, 6-Dimethoxy Phenol

The EI mass spectrum of 4-allyl-2, 6-dimethoxy phenol MW 194 is given in figure (8) the loss of [CH_3] results in M [$C_{10}H_{11}O_3$]$^{+\cdot}$ at m/z 179, while the base peak is found at m/z 91, 4-allyl-2,6-dimethoxy phenol appeared at R.T 10.983 in total ion chromatogram.

3.9. Identification of (E) -1-(3-hydroxy-2,6,6-trimethylcyclohex-1-enyl)but-2-en-1-one

The EI mass spectrum of (E) -1-(3-hydroxy-2,6,6-trimethylcyclohex-1-enyl)but-2-en-1-one MW 208 is given in figure (9) the loss of [CH_3] results in M [$C_{12}H_{17}O_2$]$^{+\cdot}$ at m/z 193, while the base peak is found at m/z 69, (E) -1-(3-hydroxy-2,6,6- trimethylcyclohex-1-enyl)but-2-en-1-one appeared at R.T 11.300 in total ion chromatogram.

3.10. Identification of of 1,1,4,4-tetramethyl-2,5-dimethylenecyclohexane

The EI mass spectrum of of 1,1,4,4- tetramethyl-2,5-dimethylenecyclohexane MW 164 is given in figure (10) the base peak is found at m/z 149 corresponding to $M[C_{11}H_{17}]^+$ due to the loss of $[CH_3]$. The 1,1,4,4-tetramethyl-2,5-dimethylenecyclohexane appeared at R.T 13.150 in total ion chromatogram.

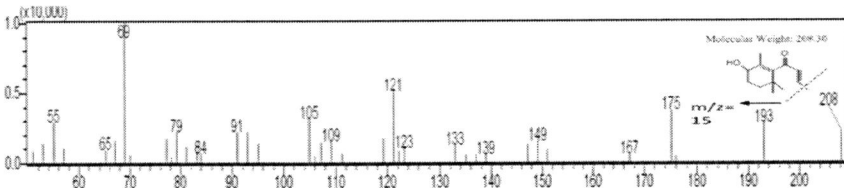

Figure 9. The mass spectrum analysis of (E) -1-(3- hydroxy-2,6,6-trimethylcyclohex-1-enyl)but-2-en-1-one.

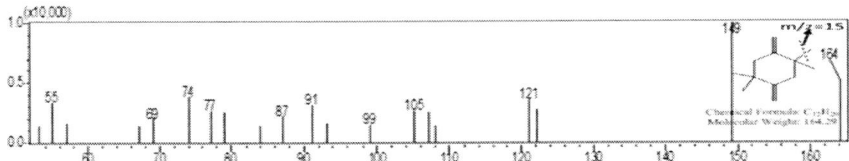

Figure 10. The mass spectrum analysis of 1,1,4,4- tetramethyl-2,5-dimethylenecyclohexane.

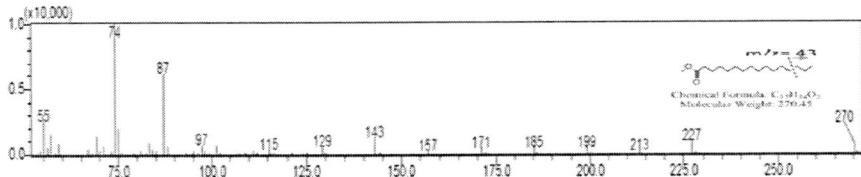

Figure 11. The mass spectrum analysis of Methyl palmitate.

3.11. Identification of Methyl Palmitate

The EI mass spectrum of Methyl palmitate. MW 270 is given in figure (11). The loss of $[C_3H_7]$· results in $[C_{14}H_{27}O_2]^{+·}$ at m/z 227. The base peak is found at m/z 74, the Methyl palmitate appeared at R.T 17.608 in total ion chromatogram.

3.12. Identification of Palmitic Acid

The EI mass spectrum of Palmitic acid .MW 256 is given in figure (12).
The loss of $[C_2H_5]$· results in $M[C_{14}H_{27}O_2]^+$ at m/z 227. While the base peak is found at m/z 73, palmitic acid appeared at R.T 18.367 in total ion chromatogram.

3.13. Identification of Methyl Stearate

The EI mass spectrum of methyl stearate MW 298 is given in figure (13) the loss of $[C_3H_7]$·results in $M[C_{16}H_{31}O_2]^+$ at m/z 255. While the base peak is found at m/z 74, methyl stearate appeared at R.T 22.050 in total ion chromatogram.

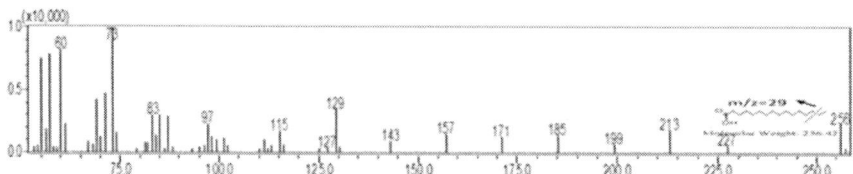

Figure 12. The mass spectrum analysis of Palmitic acid.

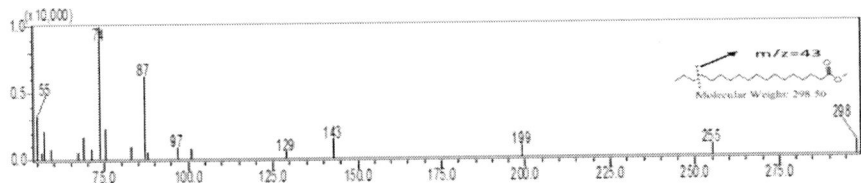

Figure 13. The mass spectrum analysis of methyl stearate.

3.14. Identification of DCM Extractable

The typical total ion chromatograms (TIC) of DCM extract were given in figure (14). While table (3) shows the compounds extracted by DCM.

Table 3. Identification and toxicity prediction of compounds extracted by DCM

R.T	Name	Structure	M.W	Area %	Class
7.608	2,6 dimethoxy phenol		154.16		phenol
R.T	Name	Structure	M.W	Area %	Class
17.667	Methyl palmitate		270.45		Esters
8.95	(3,4-dimethoxy phenoxy) trimethylsilane		226.34		phenol

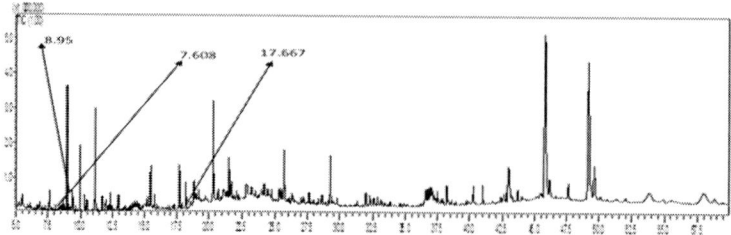

Figure 14. Total ion chromatogram of vinasse extracted by DCM. Peak numbers as R.T which is refer to table 3.

3.15. Identification of 2,6 Dimethoxy Phenol

The EI mass spectrum of 2,6 dimethoxy phenol MW 154 is given in figure (15) the base peak is found at m/z 154. The loss of [CH_3] results in M [$C_7H_7O_3$]$^+$ at m/z 139. The 2,6 dimethoxy phenol appeared at R.T 7.608 in total ion chromatogram.

3.16. Identification of Methyl Palmitate

The EI mass spectrum of methyl palmitate MW 270 is given in figure (16) the loss of [CH_3O] results in M [$C_{16}H_{31}O$]$^+$ At m/z 239, while the base peak is found at m/z 74, methyl palmitate appeared at R.T 17.66 in total ion chromatogram.

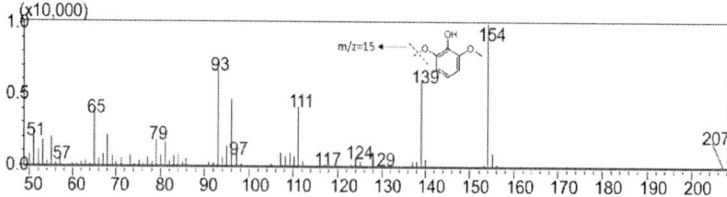

Figure 15. The mass spectrum analysis of 2,6 dimethoxyphenol.

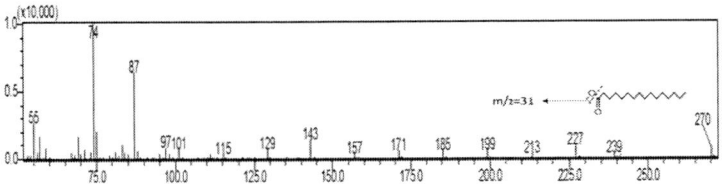

Figure 16. The mass spectrum analysis of methyl palmitate.

CONCLUSION

In this study, about 15 of organic substances identified by using GC-MS. Most compounds detected were phenolic and carboxylic acid, Hexane extraction is more effective than DCM extraction. Thus, an effective treatment method of vinasse is strongly recommended to ensure active treatment. However, it should be explained that more research is needed towards the identification of more phenolic compounds. It is believed that this procedure will solve many problems about not only the choosing of proper treatment method but also, vinasses as the by-product might be has promising antioxidant properties.

REFERENCES

Benke, M. B., Mermut, A. R. and Chatson, B. (1997). Carbon-13 CP/MAS NMR and DR-FTIR spectroscopic studies of sugarcane di stillery waste, National Research Council, Plant Biotechnology Institute.

Clesceri, L. S., Greenberg, A. E, Eaton, A. D. (1998) Standard Methods for the Examination of Water and Waste Water, 20th ed., American Public Health Association (APHA), Washington.

Cristiano, E., Rodrigues Reis and Bo Hu. (2017) Vinasse from Sugarcane Ethanol Production: Better Treatment or Better Utilization. *Journal Frontiers in Energy Research*, 5.

Decloux, M. and Bories, A. (2002). Stillage treatment in the French alcohol fermentation industry, *Int. Sugar J.*, 104, 509 - 517.

Dowd, M. K., Johansen, S. L., Cantarella, L. and Reilly, P. J. (1994). Low molecular weight organic composition of ethanol stillage from sugarcane molasses, citrus waste, and sweet whey, *Journal of Agricultural and Food Chemistry*, 42 (2): 283 - 288.

Doelsch, Emmanuel, Armand Masion, Patrick Cazevieille And Nicolas Condom. (2009). Spectroscopic characterization of organic matter of a soil and vinasse mixture during aerobic or anaerobic incubation, *Waste Management*, 29: 1929 - 1935.

Espana-Gamboa, E., Mijangos-Cortes, J., Barahona-Perez, L., Dominguez-Maldonado, J., Hernández-Zarate, G. and Alzate- Gaviria, L. (2011). Vinasses: characterization and treatments, *Waste Management and Research*, 29: 1235 - 1250.

Fitz Gibbon, F., Singh, D., McMullan, G. and Marchant, R. (1998). The effect of phenolic acids and molasses spent wash concentration on distillery wastewater remediation by fungi, *Process Biochem.*, 33: 799.

Fagier, A. Mohamed, Elmgdad. A. Ali, Kheng S. Tay and Mohamd R. B. Abas. (2016). Mineralization of organic matter from vinasse using physicochemical treatment coupled with Fe^{2+}-activated persulfate and peroxymonosulfate oxidation. *Int. J. Environ. Sci. Technol.*, 13:1189 - 1194.

Halket, John M., Daniel Waterman, Anna M. Przyborowska, Raj K. P. Patel, Paul D. Fraser and Peter M. Bramley. (2004). Chemical derivatization and mass spectral libraries in metabolic profiling by GC/MS and LC/MS/MS, *Journal of Experimental Botany*, 56: 410.

Mohana, S., Acharya, B. K. and Madamwar, D. (2009). Distillery spent wash: treatment technologies and potential applications, *Journal of Hazardous Materials*, 163: 12 - 25.

Molina-Cortés, Andrea, Tatiana Sánchez-Motta, Fabian Tobar-Tosse and Mauricio Quimbaya. (2019). Spectrophotometric estimation of total phenolic content and antioxidant capacity of molasses and vinasses generated from the sugarcane industry. *Waste and Biomass Valorization*.

Parnaudeau, V., Condom, N., Oliver, R., Cazevieille, P. and Recous, S. (2008). Vinasse organic matter quality and mineralization potential as influenced by raw material, fermentation and concentration processes, *Bioresource Technology*, 99: 1553 - 1562.

Satyawali, Y. and Balakrishnan, M. (2008). Wastewater treatment in molasses-based alcohol distilleries for COD and color removal: a review, *Journal of Environmental Management*, 86(3): 481 - 497.

Soni Tiwari, Rajeeva Gaur and Ranjan Singh. (2012). Decolorization of a recalcitrant organic compound (Melanoidin) by a novel thermotolerant yeast, Candida tropicalis RG-9, *BMC Biotechnology*, 12: 30.

INDEX

A

Abraham, 69, 86
abuse, vii, viii, 2, 4, 5, 38, 45, 85, 124
accessibility, 85
acetaldehyde, 132, 136
acetic acid, 75, 157, 159, 171, 175
acetone, 6, 23, 135, 147
acetonitrile, 9, 25, 29, 33, 34, 35, 75, 76
acid, 69, 70, 71, 73, 74, 75, 83, 131, 132, 154, 157, 160, 163, 175, 180, 181, 183, 186
acylation, 53, 59
additives, 164
adipose tissue, 39, 151, 169
adverse effects, 60
aldehydes, 163
alkylation, 44, 70
allergic asthma, 85
American Neurological Association, 98
amines, 53, 59, 64, 66
amniotic fluid, 20
amphetamines, 45, 47, 53, 54, 91, 110, 111, 114, 124, 125
antidepressant, 63, 64
antiepileptic drugs, 68
antioxidant, 133, 138, 174, 189, 190
antipsychotic drugs, 60
aqueous solutions, 176
assessment, 19, 157, 159, 170
authentication, ix, 142, 145, 146, 147, 149, 151, 154, 156, 160, 163, 164, 166, 168, 169, 170, 171, 172
authenticity, vii, viii, ix, 141, 142, 143, 144, 145, 146, 149, 153, 154, 156, 157, 160, 166, 168, 169, 170
authorities, 143, 144
autopsy, 138, 139

B

background noise, 6, 13, 58, 84
bacterial fermentation, 133
base, 177, 178, 182, 183, 184, 185, 186, 188
beverages, 132, 154, 156, 157
biological activity, 54
biological monitor, 131
biological samples, viii, 2, 5, 39, 42, 43, 44, 45, 53, 54, 56, 136

biological specimens, 2, 4, 15, 19, 21, 29, 38, 45, 47, 60, 61
biomarkers, ix, 82, 84, 142, 146, 166
blood, 5, 8, 14, 15, 17, 19, 21, 35, 38, 40, 46, 48, 51, 52, 57, 58, 60, 62, 65, 68, 71, 74, 77, 78, 79, 131, 132, 133, 137
bloodstream, 60
body fluid, 80, 135
bone marrow transplant, 71
brain, 6, 15, 21, 71
breast milk, 20, 38
breath analyzer, 129, 134

C

calibration, 12, 28, 37, 51, 79, 137
cannabinoids, 29, 38, 39, 40, 41, 42, 43, 44, 54, 55, 56, 57, 58, 59, 86, 87, 94, 95, 97, 103, 110, 111, 114, 116, 117
capillary, 6, 11, 13, 15, 61, 63, 66, 68, 70, 137, 147, 154, 160
carbamazepine, 69, 77
carbohydrates, 133
carbon, viii, 28, 38, 128, 131, 135, 136, 137, 153, 154, 156, 157, 159, 170, 171
carbon monoxide, viii, 128, 135, 136, 137
carbon monoxide (CO), viii, 128, 131, 135, 136, 137
carbon nanotubes, 28, 38
carboxylic acid, 37, 153, 174, 189
CBD, 29, 30, 31, 32, 34, 35, 36, 38, 39
central nervous system, 55
cerebrospinal fluid, 38
cheese, 160, 161, 164, 168
chemical, vii, viii, ix, 5, 41, 66, 78, 141, 142, 145, 146, 147, 149, 150, 154, 160, 165, 166, 168, 174, 175, 176
chemical characteristics, 176
chemometrics, 171
chloroform, 12, 24, 48
chromatograms, 6, 83, 178, 187

chromatographic technique, 39, 69, 85
classification, 81, 145, 149, 151, 153, 160, 166
clinical application, 85
clinical diagnosis, 82
clinical toxicology, 2, 46, 80
clozapine, 61, 62, 72
CO poisoning, 131
cocaine, 14, 15, 16, 17, 18, 19, 20, 21, 22, 23, 24, 25, 26, 27, 28, 38, 55, 87, 88, 89, 90, 91, 92, 94, 96, 101, 103, 105, 106, 110, 113, 114, 116, 117, 118, 119, 126
combustion, ix, 142, 146, 153, 156, 157, 164, 171
commercial, ix, 139, 142, 144, 156, 157, 158, 159, 164, 170
complexity, 3, 80, 143, 144, 145
composition, vii, viii, 57, 138, 141, 146, 151, 175, 190
compounds, ix, x, 3, 7, 8, 14, 15, 16, 17, 20, 27, 38, 39, 40, 43, 44, 45, 46, 52, 53, 55, 56, 57, 59, 64, 69, 78, 80, 82, 85, 128, 131, 139, 142, 144, 145, 146, 149, 153, 154, 160, 163, 164, 166, 168, 169, 170, 174, 175, 177, 178, 179, 187, 189
constituents, 163
construction, 130, 150
consumers, 46, 55, 143, 144
consumption, 14, 17, 39, 40, 54, 55, 143
contamination, 18, 41, 144
conversion reaction, 84
COOH, 39, 40, 103, 115, 120
correlation, 58, 71, 156
Correlation Optimized Warping (COW), 151
correlations, 62
cyanide, viii, 128, 131, 132, 137, 138
cytochrome, 39, 132, 133
cytochrome oxidase, 132, 133

D

data analysis, 129
data processing, 149
data set, 145, 147, 148, 149
decomposition, 69
Department of Justice, 103
derivatives, 4, 5, 7, 8, 21, 45, 46, 47, 52, 53, 54, 64, 66, 67, 69
detection, 3, 4, 12, 13, 14, 15, 16, 17, 18, 20, 28, 37, 39, 40, 42, 43, 51, 54, 57, 58, 64, 65, 68, 69, 79, 81, 83, 84, 134, 137, 144, 149, 161, 164, 175
detection system, 4, 13
detection techniques, 84
dimethylformamide, 69
discriminant analysis, 151
diseases, 85, 133, 138
distribution, ix, 142, 146, 153, 159
dizziness, 131, 132, 133
drugs, vii, viii, 2, 4, 5, 7, 8, 9, 13, 14, 15, 18, 19, 20, 38, 40, 45, 46, 52, 53, 54, 55, 58, 60, 62, 63, 64, 65, 68, 71, 72, 80, 81, 82, 85, 135, 136, 142
dynamic headspace (DH), 149, 150, 151, 169

E

economic consequences, 143
editors, 135, 138
electron, 5, 9, 10, 21, 22, 23, 24, 25, 26, 27, 28, 44, 64, 72, 121, 162, 176
electrons, 64
embolism, 138
employment, viii, 142
energy, 9, 10, 21, 22, 23, 24, 25, 26, 27, 28, 72, 75, 76, 77, 79, 140, 162
enzyme-linked immunosorbent assay, 18
ester, 14, 17, 18, 19, 20, 24, 25, 26, 27, 69

ethanol, vii, viii, ix, 4, 14, 128, 132, 154, 156, 158, 170, 173, 174, 175, 176, 178, 179, 182, 189, 190
ethers, 160
ethyl acetate, 31, 32, 34, 36, 42, 47, 50, 73, 76, 77, 153, 154
ethylbenzene, 133, 134
euphoria, 71
European Monitoring Centre for Drugs and Drug Addiction (EMCDDA), 54, 55, 92, 95
evolution, 54, 60
experimental condition, 7, 149
exposure, 3, 17, 20, 52, 133, 144
extraction, x, 3, 10, 11, 12, 15, 16, 17, 18, 19, 24, 28, 37, 38, 42, 51, 53, 54, 58, 60, 61, 63, 68, 71, 79, 80, 85, 144, 147, 149, 151, 153, 154, 156, 174, 176, 189
extracts, ix, 13, 19, 84, 142, 153, 154, 159

F

false positive, 18
families, ix, 142, 146
fat, 83, 151, 160, 161, 171
fatty acids, ix, 69, 83, 142, 146, 160
fermentation, 133, 154, 156, 157, 190, 191
filament, 9, 10, 22, 72
film thickness, 9, 10, 11, 12, 21, 22, 23, 24, 25, 26, 27, 28, 29, 72, 73, 74, 75, 76, 77, 78, 79, 160, 165, 176
fingerprinting, ix, 142, 145, 146, 147, 152, 153, 163, 166, 168
fingerprints, 152, 153
fire-related death, 131
fires, 131
flame, 4, 13, 43, 54, 64, 108, 109, 132, 133, 137, 139, 160, 171
flame photometric detector (FPD), 133, 139
flame thermo-ionic detector (FTD), 13, 108, 109, 132, 137

flavor, 155, 164
fluid, viii, 2, 5, 14, 19, 20, 26, 27, 29, 33, 40, 46, 58, 60, 62, 66, 70, 72, 76, 78, 80, 139
fluorescence, 140, 145
food, vii, viii, ix, 40, 141, 142, 143, 144, 145, 146, 149, 151, 154, 156, 159, 160, 163, 166, 167, 168, 169, 171, 172
food adulteration, vii, viii, 141, 143, 144
food authentication, ix, 142, 146, 151, 163, 166, 168, 169
food chain, 143, 144
food products, 142, 144, 145, 146, 149, 156, 160, 166
food safety, vii, viii, 141, 142, 144
forensic, v, vii, viii, 1, 2, 4, 9, 14, 16, 18, 20, 38, 39, 43, 45, 46, 52, 53, 56, 59, 63, 80, 85, 87, 89, 90, 93, 94, 101, 102, 103, 104, 106, 108, 110, 111, 113, 114, 116, 117, 119, 120, 126, 127, 128, 129, 131, 132, 133, 134, 135, 136, 137, 138, 139
forensic diagnosis, vii, viii, 128, 132
forensic toxicology, vii, viii, 2, 4, 14, 45, 59, 85
fragments, 43, 150
fraud, viii, 142, 143, 145
fruits, ix, 142
functional food, 143

G

gas chromatography coupled to mass spectrometry (GC-MS), vii, ix, 5, 7, 8, 12, 15, 17, 18, 19, 20, 21, 22, 27, 28, 29, 30, 31, 32, 33, 34, 35, 36, 38, 40, 41, 43, 46, 47, 48, 49, 50, 51, 53, 56, 57, 59, 60, 61, 62, 63, 65, 66, 67, 68, 70, 71, 72, 73, 74, 75, 76, 77, 79, 81, 83, 85, 87, 97, 100, 105, 111, 115, 117, 119, 142, 146, 147, 148, 149, 150, 151, 152, 160, 163, 166, 168, 169, 171, 173, 174, 175, 176, 177, 189
gas sensors, 136
gaseous or volatile substances, vii, viii, 128, 134
gastric mucosa, 82
gastrointestinal diseases, 133, 138
GC-FID, 43, 54, 64, 160
GC-flame ionization detector (GC-FID), 43, 54, 64, 160
genetic defect, 83
geographical origin, 155, 164, 165
glycerol, 163, 175
glycine, 79
growth, 151, 156
growth rate, 151

H

H_2, 133
H_2S, 133
hair, viii, 2, 5, 6, 13, 14, 17, 18, 19, 40, 49, 51, 52, 58, 65
height, 128
helium, 6, 66, 128
hemoglobin, 131
hepatocellular carcinoma, 82
hepatotoxicity, 82
hexane, x, 31, 32, 34, 35, 36, 42, 73, 174, 177, 178, 179
human, vii, viii, 6, 13, 15, 17, 18, 41, 55, 61, 65, 66, 82, 135, 136, 137, 141, 144, 172
human exposure, 15
human health, vii, viii, 141, 144, 172
hybrid, 85
hydrogen, viii, 6, 116, 128, 131, 132, 133, 135, 138, 139, 154, 156, 157, 170
hydrogen cyanide, 131, 132
hydrogen gas (H_2), 133, 138
Hydrogen Sulfide (H_2S), 133, 139
hydrolysis, 5, 30, 31, 34, 42

I

identification, ix, 5, 8, 14, 18, 43, 45, 52, 53, 56, 57, 65, 80, 81, 85, 128, 142, 144, 145, 146, 163, 166, 174, 177, 189
illicit drug market, 55
images, 115
improvements, 3, 4
in vivo, 41, 131, 138
incomplete combustion, 131
individuals, 19
infrared analyzer, 132
ingestion, 132, 134, 139, 140
inhibition, 132, 133
insecticide, 140
integrity, 65, 143, 144, 145, 146, 149, 160, 166
interface, 11, 12, 27, 50, 74, 75, 79, 157
ionization, 4, 5, 41, 43, 44, 51, 54, 57, 64, 66, 75, 76, 77, 78, 79, 81, 160, 171
ions, 5, 7, 14, 19, 57, 67
isotope, ix, 66, 83, 142, 146, 153, 154, 156, 157, 159, 164, 169, 170, 171, 182
isotope ratio mass spectrometry (IRMS), ix, 142, 146, 153, 154, 156, 157, 158, 159, 164, 166, 169, 170, 171, 172
issues, ix, 142, 166

K

ketones, 53
kidney, 21

L

labeling, 143
lack of control, 156
lactic acid, 175
lactose intolerance, 133
law enforcement, 144

LC-MS, 56, 57, 60, 81, 89, 97
LC-MS/MS, 60, 89
lead, 3, 63, 81, 84, 133
legislation, 55
leukemia, 71
lignin, 175
linear discriminant analysis (LDA), 151
liquid chromatography, vii, 2, 3, 4, 39
Liquid Chromatography-Tandem Mass Spectrometry, 105
liquid phase, 18, 37, 42, 151
liver, 15, 38, 58
loss of consciousness, 133
low temperatures, 52

M

majority, 80
malabsorption, 133
malignant tissues, 82
malodor, 133, 135
management, 3
manipulation, 143
mass spectrometry, vii, ix, 2, 3, 5, 7, 8, 12, 14, 15, 17, 18, 19, 20, 21, 22, 27, 28, 29, 30, 31, 32, 33, 34, 35, 36, 38, 40, 41, 43, 46, 47, 48, 49, 50, 51, 53, 56, 57, 59, 60, 61, 62, 63, 65, 66, 67, 68, 70, 71, 72, 73, 74, 75, 76, 77, 79, 81, 83, 85, 87, 97, 100, 105, 111, 115, 117, 119, 140, 142, 146, 147, 148, 149, 150, 151, 152, 153, 157, 160, 161, 163, 164, 166, 168, 169, 170, 171, 173, 174, 175, 176, 177, 189
matrix, 5, 15, 19, 40, 52, 57, 58, 71, 79
measurement, ix, 66, 129, 131, 142, 146, 164
measurements, 60, 63, 64, 65, 132, 157, 170, 175
meconium, 41, 49, 52
medicine, v, 87, 92, 97, 99, 101, 106, 120, 126, 127, 129, 134, 136, 137, 138

membranes, 160
Metabolic, 97, 100, 102, 113, 121, 125
metabolic pathways, 166
metabolism, 40, 80, 83, 154, 172
metabolites, ix, 6, 8, 14, 15, 17, 18, 19, 20, 21, 38, 39, 56, 58, 60, 65, 67, 70, 80, 81, 82, 142, 146, 149, 155, 166
metabolome, 80, 81, 82, 83
metabolomics, 80, 81, 83, 84, 85, 87, 88, 92, 94, 97, 101, 102, 103, 106, 118, 119, 122, 123, 125, 152, 166
metabolomics fingerprinting, 152, 166
metastasis, 82
methamphetamine, 46, 48, 52, 53
Methamphetamine, 47, 48, 49, 50, 93, 101
methanol, 9, 10, 11, 24, 25, 48
methodology, 52, 53, 163, 164, 165
microorganisms, 174
mineralization, 175, 191
molecular weight, 14, 53, 175, 190
molecules, 137
morphine, 5, 7, 8, 9, 10, 11, 12, 13
multidimensional, ix, 142, 164, 166
multidimensional gas chromatography (GC×GC), ix, 142, 146, 163, 164, 172
multiplier, 24
multivariate analysis, 168

N

National Academy of Sciences, 121
National Research Council, 189
natural food, 143
nausea, 131, 133
neurotoxicity, 82
New England, 101
new psychoactive substances, vii, viii, 2, 4
nitrogen, 10, 13, 61, 64, 65, 66, 69, 70, 128, 131, 132, 137, 154, 176
nitrogen phosphorus detector (NPD), 10, 13, 61, 64, 65, 66, 69, 70, 132

nitrosamines, 68
non-target analysis, 142
non-targeted approaches, 145, 166
non-targeted screening, 152
NPS, 54, 55, 56, 57, 58, 59, 95, 113, 122
nuclear magnetic resonance, 81

O

oil, 151, 162, 163, 165
olive oil, ix, 142, 143, 164
opiates, 4, 5, 6, 7, 8, 13, 14, 18, 88, 90, 100, 103, 106, 114, 115, 116, 117, 119, 124
opioids, 15, 21, 54
organic compounds, vii, ix, x, 173, 174
organic matter, 174, 175, 190, 191
organic solvents, 38
oxidation, 154, 190
oximetry, 131
oxygen, 131, 138, 174

P

partial least squares regression, 145, 153
partial least squares regression-discriminant analysis (PLS -DA), 145, 153
partition, 53
pasture, 149, 151, 152
pathology, 136
pathway, 154
pathways, 57, 80, 159, 170
pattern recognition, 147, 160
penalties, 56
pesticide, 119, 134, 139
pH, 11, 12, 75
pharmaceutical, 64
pharmaceuticals, 135
pharmacokinetics, 68
phenol, x, 174, 178, 179, 180, 182, 183, 184, 187, 188
phenolic compounds, x, 174, 175, 189

phenytoin, 68, 75, 77
phosphate, 22, 24, 75, 77
phosphoenolpyruvate, 171
phosphorus, 10, 13, 61, 64, 65, 66, 69, 70, 90, 132, 137
placenta, 20, 29, 41
plants, 147, 154, 157, 159, 166, 171
plasma levels, 61, 71
PLS, 145, 153
polar, 7, 53
polarity, 13, 19, 53, 144, 164
pollution, 174
polydimethylsiloxane, 50
polymer, 165
polymorphism, 132
polymorphisms, 136
polypropylene, 30
population, 85
portability, 130
Portugal, 1, 2
potassium, 22, 131
precipitation, 12, 37, 42, 154
pregnancy, 41
preparation, iv, x, 14, 42, 46, 54, 57, 58, 61, 63, 71, 131, 151, 174
preservation, 152, 168
prevention, ix, 142, 146
principal component analysis, 111, 145, 148, 149, 151, 152, 153, 165
principal component analysis (PCA), 111, 145, 148, 149, 151, 152, 153, 165
principles, 153
producers, 143
profitability, 174
prognosis, 3
propionic anhydride, 7
protected designations of origin (PDO), 143, 161, 162
protection, 167
proteins, 144
psychosis, 71
public health, vii, ix, 45, 54, 55, 173

purification, 3, 61, 85, 157
purity, 39, 55, 176
PVP, 102
pyrolysis, 154, 156

Q

quality control, 55, 144, 171
quantification, 3, 7, 15, 16, 17, 18, 39, 40, 41, 42, 46, 53, 56, 60, 62, 65, 66, 68, 71, 72, 80, 81, 83, 85, 86, 88, 89, 90, 100, 101, 102, 103, 106, 110, 114, 117, 123, 125, 128, 171
quantitation, 8, 12, 28, 37, 51, 79, 87, 88, 91, 100, 109, 110, 118, 122, 131, 132, 145, 160, 161
quetiapine, 62, 72

R

radicals, 138
ramp, 5, 9, 10, 11, 21, 27, 30, 48, 50, 147
ratio analysis, 156, 157, 170
raw materials, 157, 159
reactions, 39, 44, 59
reagents, 16, 53, 55, 56, 64, 84
recommendations, iv
recovery, 16, 42
recreational, 55
regulatory bodies, 143
relevance, vii, 2, 14, 16, 61, 151
remediation, 190
requirements, 63, 167
researchers, 3, 6, 16, 80, 175
resistance, 129
resolution, viii, 2, 3, 15, 59, 63, 64, 84, 152, 169
resources, 46
respiration, 132
respiratory failure, 133
response, 81

S

safety, 124, 143, 149, 167
salts, 55, 131
sanctions, 56
saturated fat, 83
saturated fatty acids, 83
savings, 156
scope, ix, 142, 146, 153
selective serotonin reuptake inhibitor, 66
selectivity, 5, 13, 38, 53, 57, 59, 63
semiconductor, vii, viii, 128, 129, 135, 136
semiconductor gas sensor, vii, viii, 128, 129, 136
semiconductor sensor, 129
sensitivity, vii, viii, 3, 5, 6, 7, 8, 13, 38, 42, 43, 53, 57, 59, 60, 62, 63, 64, 69, 70, 84, 85, 128, 131, 132, 133
sensor, vii, viii, 128, 129, 130, 135, 136, 137, 138
sensor gas chromatography (sGC), vii, viii, 128, 129, 130, 131, 132, 133, 134, 137, 138
septum, 14
sertraline, 66, 67, 77
serum, 5, 15, 21, 27, 40, 65, 70, 71, 78, 83
showing, 40, 155, 163
silica, 6, 9, 33, 43, 66, 73, 75, 76, 79
sodium, 11, 12, 68, 73, 79, 131
sodium hydroxide, 79
solid phase, 10, 15, 18, 54, 171
solution, 22, 73, 76, 80, 154
solvents, 38, 42, 44, 46, 58, 81, 134, 176
species, 64, 154, 156, 164
spectrophotometric method, 136
spectrophotometry, 131
spectroscopy, 81, 145

risk, vii, ix, 41, 54, 55, 173
risk assessment, 55
room temperature, 53

spleen, 84
stability, 7, 8, 44, 58
standard deviation, 158
state, 63, 80, 136, 156, 168
static head-space (HS), 24, 28, 31, 37, 42, 50, 51, 79, 97, 120, 129, 131, 132, 134, 140, 152, 156, 157, 158, 159, 163, 164, 168, 171
structure, 13, 44, 54, 56, 57, 59, 66, 177
sugarcane, 174, 175, 189, 190
sugarcane vinasse, 174, 175
sulfuric acid, 132, 175
supervision, 19
sweat, viii, 2, 41, 50, 52
symptoms, 19, 131, 132
synthesis, 154, 157

T

tachypnea, 132
target, 13, 81, 82, 129, 142, 163
fingerprinting, 142
targeted approaches, 144, 160
targeted GC-MS approaches, 146, 160
techniques, viii, ix, 2, 3, 15, 19, 38, 42, 46, 56, 60, 61, 69, 70, 71, 81, 85, 135, 142, 145, 146, 149, 151, 159, 160, 163, 166, 175
temperature, 6, 9, 10, 11, 12, 21, 22, 23, 24, 25, 26, 27, 28, 29, 30, 31, 32, 47, 48, 49, 50, 51, 72, 73, 74, 75, 76, 77, 78, 79, 147, 149, 154, 157, 163, 177
therapeutic drug monitoring, 60
thermal conductivity detector (TCD), 131
thermal stability, 14
tissue, 12, 20, 82, 84, 139, 150, 151
toluene, 133, 134, 139
toxic substances, 131, 143
toxicity, 82, 139, 179, 187
toxicology, v, vii, viii, 1, 2, 4, 5, 14, 45, 46, 59, 60, 80, 85, 88, 89, 90, 91, 92, 93, 95,

Index 201

97, 98, 99, 100, 101, 103, 104, 105, 106, 108, 109, 110, 111, 113, 114, 115, 116, 117, 118, 119, 120, 121, 122, 123, 124, 136
training, 148, 149
transformation, 64
transplantation, 133
treatment, ix, 42, 45, 52, 53, 57, 61, 63, 71, 81, 83, 131, 136, 144, 151, 173, 175, 189, 190, 191
tricyclic antidepressant, 63
tumor, 82

U

umbilical cord, 20
underlying mechanisms, 82
uniform, 67
urine, 5, 8, 13, 14, 15, 16, 17, 35, 38, 40, 45, 46, 48, 49, 51, 52, 53, 58, 62, 63, 67, 77, 80, 82, 83, 132

V

vacuum, 89
validation, 38, 149
valorization, 174
varieties, 152, 168
vegetable oil, 151

venlafaxine, 67, 77
versatility, viii, 2
vinasse, vii, ix, 173, 174, 175, 176, 177, 178, 188, 189, 190
volatile organic compounds, 149, 151
volatile organic compounds (VOCs), 149, 151
volatility, 14, 44, 53, 62
volatolome, ix, 142, 146, 151, 152, 166
volatolomics, 149, 151, 152, 166, 168, 169

W

waste, 174, 189, 190
wastewater, ix, 173, 174, 190
water, ix, 23, 25, 53, 135, 142, 154, 157, 174, 176
World Health Organization, 55, 124
worldwide, 3, 14, 38, 54, 63, 143

X

xylene, 133, 134, 140

Y

yeast, 191